THE JOURNAL OF

JAMES MACRAE

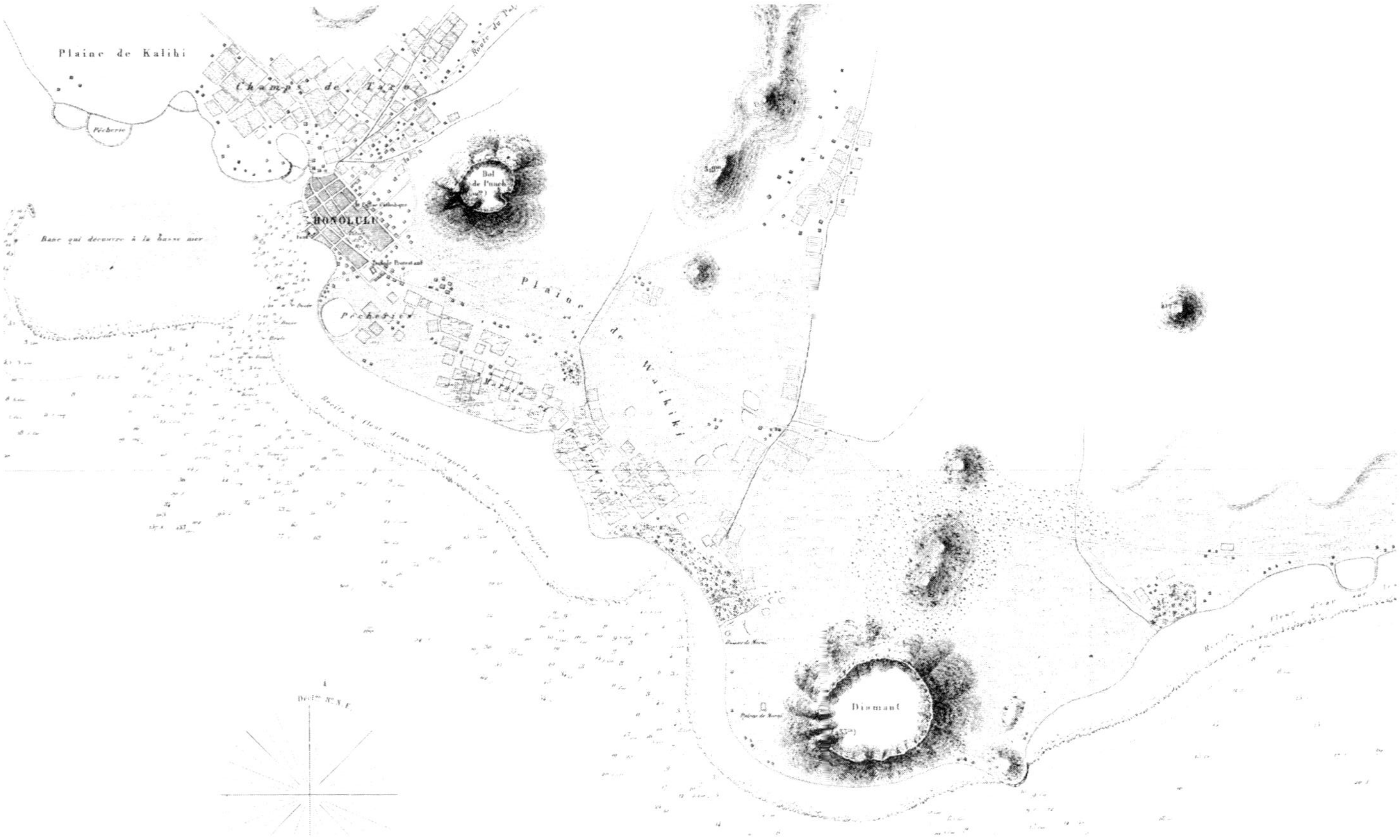

Detail of 1855 "Plan du Mouillage d'Honolulu" by Joseph Marie Henri de La Passe, lieutenant aboard the *Eurydice*. The map is courtesy of Hamilton Library, the University of Hawai'i, and Gaylord Wilcox, Editions Limited; the original is in the Library of Congress.

THE JOURNAL OF JAMES MACRAE

Botanist at the Sandwich Islands, 1825

Edited with commentary by

BRIAN RICHARDSON

North Beach-West Maui Benefit Fund Inc.

Lahaina, Maui, Hawaiʻi

Printed in the United States of America

24 23 22 21 20 19 6 5 4 3 2 1

ISBN 978-0-8248-8315-7 (pbk : alk. paper)

Also includes: *With Lord Byron at the Sandwich Islands in 1825*
Being Extracts from the MS Diary of James Macrae, Scottish Botanist

Originally published by William Wilson in Honolulu, 1922

Published by the North Beach-West Maui Benefit Fund, Inc.
P O Box 11329
Lahaina, Hawai'i 96761

Distributed by University of Hawai'i Press
2840 Kolowalu Street
Honolulu, HI 96822-1888

Print-ready files provided by North Beach-West Maui Benefit Fund, Inc.

Cover illustration: "View at Lahaina" from Maria Callcott's *Voyage of H. M. S.* Blonde *to the Sandwich Islands, in the years 1824-1825. Captain the Right Hon. Lord Byron, commander.* London, J. Murray, 1826.

This book is dedicated to my mother,
Mary Richardson.

Contents

List of Illustrations

Foreword

Publishing the journals of James Macrae, a botanist attached to the diplomatic voyage of Lord Byron returning King Liholiho and Queen Kamāmalu following their deaths in England, provides some of the first significant European descriptions of the people, when Lahaina was still the capital of the Kingdom—some 50 years after Cook first visited Hawai'i. Although bowdlerized selections from the journals were printed some hundred years ago and reprinted nearly fifty years ago, the material is largely unavailable to the West Maui community and the general reading public.

The value of this book is that it presents Macrae's journals while he was in Hawai'i with their unpolished, thick descriptions of Macrae's observations together with Wilson's 1922 bowdlerized extracts. Brian Richardson then goes through and provides context and analysis regarding the journals from the perspective of present-day historiography as well as tries to better understand the meaning of the editorial decisions that constituted Wilson's 1922 extracts.

This project has occurred simultaneously with another project by U'ilani Tanigawa-Lum and Keely Kau'ilani Rivera involving the indexing and translating of a selection of newspaper articles on Hawaiian life in the nineteenth century in West Maui—although these stories emerged some forty or more years later. These two projects represent efforts to provide greater access to more material regarding West Maui in earlier times, particularly the nineteenth century.

Among the various threads important today to our understanding of West Maui and Hawai'i in the 1820s, Macrae's observations provide valuable information regarding the final chapter in Liholiho's and Kamāmalu's tour around the world. The missionaries were only beginning their ascent as Ka'ahumanu and her group began to institute sumptuary laws based upon Calvinist morality at the same time as Macrae made his observations. His observations also trace various relations and connections of a global political and economic totality within which Hawai'i was just beginning to become entwined.

Lance D. Collins
Hā'enanui, Honokōwai

Acknowledgments

The editor would like to thank Lance Collins for his key support, Katie Fox for her amazing editorial help, Colleen Fox, Tom Spreen, and Richard Fulton for their extensive comments and suggestions, Lucille Aono and Wanda China for their help with the production, and Kalani Meinecke, Emalia Keohokalole, Elizabeth Seaton, Dore Minatodani, Jodie Mattos, Erica Guillen, Cai Parryjones, Crestina Forcina, Ted Kwok, Gary Fitzpatrick, and Gaylord Wilcox for their assistance with accessing, preparing, and understanding the materials.

Notes on the Text

This publication combines two distinct works: first, a transcript of James Macrae's original journal written while in the Hawaiian Islands in 1825, created with permission from the Royal Horticultural Society in London, and second, a reprint of William Wilson's *Extract from the Journal of James Macrae,* published in 1922.

For the transcript of the original journal, spelling has been updated and standardized, with the exception of Hawaiian terms where Macrae's original spelling is typically maintained. Abbreviations have been expanded and some punctuation has been added to increase readability. "Tho." has been rendered "thermometer," which is the spelling that Macrae uses in his journal. For the most part, Macrae's grammar has been maintained, even if it does not conform to current practices. Thus the sentence:

> Along the coast on both sides of Diamend hill are number of the natives huts pleasently situated underneath the shade of Banana & coco-nut trees, but in the ravines towards the interior (which are cheifly of Volcanic origin) they are thinly scattered at the botom of the ridges on each side & without Coco-nut trees which are seldom seen to grow far from the sea coast.

Is rendered:

> Along the coast on both sides of Diamond Hill are [a] number of the natives' huts pleasantly situated underneath the shade of banana & coconut trees, but in the ravines towards the interior (which are chiefly of volcanic origin) they are thinly scattered at the bottom of the ridges on each side & without coco-nut trees which are seldom seen to grow far from the sea coast. (page 65)

The result is that the transcript of Macrae's journal is sometimes not easy to read, but it is more faithful to what he originally wrote.

Minor editorial changes have been made to the reprint of Wilson's *Extract from the Journal of James Macrae*. Obvious typographical errors have been fixed. "Bloxom" has been changed to "Bloxam" throughout. Errors in transcription, some of which are discussed in the Afterword, have been left as they occurred in print. The locations of footnotes and images have been modified to fit the new pagination.

Throughout the literature, Macrae's name is variously spelled Macrae, McRea, McCrae, McRae, and M'Rea. In this publication, "Macrae" will be used.

View Near Honoruru, Oahu. From Maria Callcott's *Voyage of H. M. S.* Blonde *to the Sandwich Islands, in the years 1824–1825. Captain the Right Hon. Lord Byron, commander.*

INTRODUCTION

James Macrae's journal was written while he served as botanist aboard HMS *Blonde,* a 46-gun frigate that was sent to the Pacific Ocean in 1824–1825. The ship sailed from Britain to Brazil and then around Cape Horn to Chile. During this historical period, there were many British ships in the Pacific, including a significant British naval presence in Valparaiso protecting expanding British economic and political interests. The *Blonde* then sailed to the Galapagos Islands, the Hawaiian Islands, and finally back to Chile, St. Helena, and Britain.

Prior to being assigned to the *Blonde*, Macrae had worked as a gardener at the Botanical Garden of St. Vincent in the Caribbean. Macrae was not an aristocrat, and likely had no wealth, but he belonged to a group of educated scientists in Britain that arose during the Scottish Enlightenment of the late 18th and early 19th century. Under the employment of the Royal Horticultural Society of London, Macrae became the official botanist aboard the *Blonde*, serving alongside other supernumeraries such as the artist Robert Dampier, the naturalist Andrew Bloxam, and his brother Rowland Bloxam, the chaplain. He was entrusted with fruit trees and other plants to be transplanted in the Hawaiian Islands; he was also expected to analyze the flora and collect specimens wherever he went. During the trip, Macrae wrote a journal, directed to his patrons at the Society, that detailed his activities and observations. He was specifically concerned with the plants that were encountered along the way, but people, places, and events were also included, and often described with a similar attention to detail. After returning to Britain, Macrae was given a post at the Peradeniya Botanic Garden in Ceylon (Sri Lanka), where he died in 1830. His journal of the voyage on HMS *Blonde* has been stored in the Society's library since the voyage ended and has yet to be published in full.

The *Blonde's* captain, Lord George Anson Byron, cousin to the poet George Gordon Byron, was a career naval officer who had served during the Napoleonic Wars. The *Blonde* was one of the first large European warships to reach the Hawaiian Islands. Many ships of European design had already arrived in the Hawaiian Islands, including some warships. The ships captained by James Cook and George Vancouver, while technically part of the Royal Navy, were colliers pressed into naval service. The *Blonde*, on the other hand, was a 46-gun frigate that, at a length of 155 feet, was 50 feet longer than Cook's *Resolution* and over twice the displacement. Its sheer size drew attention—according to the account in the journal of James Macrae.

> What most attracted the natives' notice was the unusual large size of our ship to them, which whenever they would come on deck they began sometimes to the number of half a dozen to fathom her length from the stern to the bow along the bulwark & at other times laying themselves flat on the quarter deck. (page 41)

European ships were common in the Hawaiian Islands by the 1820s. The *Blonde's* stop in the Hawaiian Islands was notable for returning the bodies of King Kamehameha II (Liholiho) and Queen Kamāmalu, both of whom had died of measles while visiting Britain in 1824. The Hawaiian royalty had been in England in part to meet with King George IV to discuss the possibility of Britain establishing a protectorate over the islands (Dampier and Joerger vii). They died before they were able to meet with the British king.

When the *Blonde* arrived off Hawai'i Island in May 1825, the Hawaiian Islands were undergoing profound change. Kamehameha I had died in 1819. The *kapu* system had been challenged by Liholiho later that year, with major resistance to the change ending with the battle of Kuamo'o on Hawai'i Island in December. A European presence in the islands was already established by this time. The British sailors Isaac Davis and John Young, as well as the Spaniard Francisco de Paula Marin, had become important European advisors to the Hawaiian *ali'i* (chiefs) beginning in the 1790s. American missionaries first arrived in 1820 and were exerting increased influence over people in the islands. Finally, the first official European consul, Mr. Richard Charlton from Britain, arrived two weeks before the *Blonde* to establish a diplomatic presence. While in the Hawaiian Islands, the *Blonde's* officers, gentlemen, and some of its crew participated in the royal funeral, explored parts of the island chain, collected specimens and artifacts, and added entries in their journals. Robert Dampier, the expedition's artist, also sketched and painted people, landscapes, and objects.

A comprehensive history of the publications connected to the voyage of the *Blonde* remains to be done. The current transcription of Macrae's journal will add to the available literature by building on, and in part correcting, William Wilson's 1922 publication, *With Lord Byron at the Sandwich Islands in 1825*.

Prior to Wilson's extract of Macrae's journal, there were two primary books that detailed the *Blonde's* stay in the islands: Maria Callcott created an account from the perspective of the people on the ship, while Charles Stewart wrote from the perspective of an American missionary living in Honolulu.

In 1826, Lady Maria Callcott's *Voyage of H.M.S.* Blonde *to the Sandwich Islands, in the Years 1824–1825. Captain the Right Hon. Lord Byron, Commander* was published by John Murray in London (Barnes and Barnes 53) with a print run of 750 copies. Callcott had never been to the Hawaiian Islands. She spent some time in Chile and included a detailed account of the country's 1822 earthquake in her *Journal of a Residence in Chile* and later in *Transactions of the Geological Society of London*. In the official publication of Byron's voyage, Callcott was not credited as an author. Instead, the print narrative was "compiled from the various journals and notes made by some of the officers and other gentlemen that accompanied Lord Byron on his interesting voyage to the Sandwich Islands" (Callcott iii). Callcott notes with regret that Mr. Bloxam "was prevented by the suddenness of his departure from England, in order to fulfil his duty in a distant colony, from arranging his own papers and those of his companions." (Callcott iii). Bloxam had travelled to Australia immediately after returning to Britain on board the *Blonde* and his journals would not be published until 1925, by the Bernice P. Bishop Museum Press. Selections from Callcott's account would be reprinted in other works, such as in Frederick Beechey's account of his own Pacific voyage from 1825 to 1828, and hers has remained the primary source for an account of the complete voyage.

While mentioning Dampier and the brothers Bloxam, Callcott makes no mention of James Macrae by name. She writes in the preface:

> It is to be regretted, that the practised collector of botanical specimens who went in the *Blonde* to the Sandwich Islands should not have furnished any account of the plants, useful or curious, which he collected for

> the Horticultural Society, or that some of the very enlightened members of that Society should not have done so. The field is in some respects new; and it is acknowledged by all the foreign navigators, that the collection made during the *Blonde's* voyage is one of the most curious in Europe. The few notices that are to be found in the present work concerning the useful plants were taken chiefly from the books, printed and manuscript, in Sir Joseph Banks's library, to which the kindness and liberality of Mr. Brown allowed the Editor free access. (page 137)

Macrae's journal, including account of the plants, can be found in the archives of the Royal Horticultural Society in London, his employer during the voyage. Because Callcott was not able to access the document, however, Macrae's journal was not integrated into the account of the voyage published at the time.

In 1828, Charles Samuel Stewart, an American Presbyterian in the islands at the time, published his *Journal of a Residence in the Sandwich Islands During the Years 1823, 1824, and 1825*. Prior to the *Blonde's* arrival, he had been living in the Hawaiian Islands for just over two years and was actively promoting Christianity as part of an early missionary community. The events related from Lord Byron's visit included observations of the royal funeral in Honolulu, sailing to Hawai'i Island on the *Blonde*, and climbing Kilauea. The two and a half-month period of Byron's visit accounts for the last tenth of his book. Soon after the *Blonde* left Hawai'i, Stewart returned to the United States on account of his wife's illness.

Before turning to consider Macrae's journal and the extract produced by William Wilson, it is worth comparing accounts of the voyage that were available in print before the 1920s. A broader analysis of the alternate descriptions would also include Hawaiian-language sources, European and American archival sources, and publications after 1920s, including transcriptions of the journals written by Robert Dampier and Andrew Bloxom. To illustrate the differences between Callcott's and Stewart's accounts of the *Blonde's* stay in the islands, which were both polished for the printing presses of London and New York, here is a comparison of the respective accounts of the funeral of the King and Queen held in Honolulu on May 11, 1825. Footnotes have not been included. Vertical space has been added to place the passages closer together, and key phrases have been bolded at points to help align the narratives.

Callcott *Voyage of H.M.S.* **Blonde** *to the Sandwich Islands* **pages 124 to 130**	**Stewart** *Journal of a Residence in the Sandwich Islands* **pages 149 to 154**
May 11.—This day the bodies of King Tamehameha the Second, and his Queen Kahamalu, were removed from His Majesty's Ship Blonde, and deposited, with Christian rites, in a house in the capital of their own native dominions, which house is to serve as their sepulchre until a proper tomb can be constructed for them. An event so singular in the history of the world as this is, will deserve every detail which can be given: but first we may consider the extraordinary fate of these young people. Born in a country which had been for ages concealed by the ocean from the rest of the world, and which had only, in the reign of their grandfather, been made known to the civilized part of mankind, they were nurtured among the ignorance and superstitions of barbarism itself. The transcendant qualities of their father, a conqueror and legislator, had alone opened to their country a prospect of rising to a station among the cultivated nations of the earth. Yet young as they were, untrained by scholarship or example, they had broken down the barriers of superstition, paved the way for laws and true religion, introduced letters, and, in hopes of benefiting their country, and securing the alliance and protection of the state which they esteemed most likely and most able to guard them, yet leave them free to improve, and not oppress them, they had undertaken no less a voyage than half the nation, yet anxious for their distant people, and grieving that they had only accomplished the object of their heroic	*(Stewart's account begins below)*

expedition. Perhaps the perfect faith reposed in the English by the people of the Islands, is the strongest proof that ever could be given by a whole nation of simple-mindedness and freedom from guile. There was not a moment's irritation, not a moment's suspicion that unfair means had been used to shorten their days; and we were received as brothers who would sympathise with their grief, and as friends who would be glad to heal their wounds.	
At 11 A.M. the bodies were placed in the frigate's launch, and carried in procession to the shore, while the colours were hoisted half-mast high, and **minute guns were fired from the ship.** The procession itself was arranged as follows: First, the launch bearing the coffins. The captain's gig, with himself and some of his officers. The barge, with the marines and the band. Pinnace and cutter, with the rest of the officers and midshipmen. Each boat had its flag only half hoisted. As the boats approached the shore, the whole of the chiefs, with the exception of Karaimoku, whose ill health rendered his attendance impossible, came to the beach to receive the remains of their sovereigns. Custom forbids the attendance of the common people on such occasions; accordingly, none were immediately present, though, on either side of the road, without the line formed by the guard, they were assembled in multitudes; for the late sovereigns had been peculiarly beloved by the common people.	*Wednesday*, 11. Suitable arrangements having been previously made for the landing of the bodies of the king and queen, at **11 o'clock this morning**, the **minute-guns of the frigate,** with a procession of barges moving from her, became signals to us to be in readiness on the point, to take the places assigned to us in the procession which was to escort the remains of our friends and pupils to a temporary mausoleum. All the chiefs, except Mr. Pitt, who was too much indisposed to have endured the fatigue, with every foreigner of respectability in the neighbourhood of Honoruru, were on the ground some time before the barges reached the shore. Mr. Charlton improved the period in forming them in the order in which they were to walk. At twelve o'clock, the procession began to move through a **double line of native soldiers,** formed on each side of the street, from the fort to the chapel, a distance of near half a mile, in the following manner:—

Two cars had been prepared to receive the coffins: they were of the form usual for carriages used by the women of high rank, and covered with black tapa; canopies and festoons of the same material were raised over them, and they were each drawn by eight chiefs. The coffins themselves excited great admiration; they were covered with crimson velvet, and ornamented with silver gilt plates and handles. On the king's coffin was the following inscription in Hawaiian and English: Tamehameha II. Elii no nahina o Awaii make i Pelikani 28 Makaiki Kaik i ke mahoe neua o Kemakaihi 1824. Moa ino no Komakou Elii Iolani. Tamehameha II., king of the Sandwich Islands, died, 14th July 1824, in London, in the 28th year of his age. May we ever remember our beloved king Iolani. On the queen's there was nearly the same: Tamehamelu Elii no nahaina o Awaii, make i Pelikani 22 Makaiki kaiku, London, 8 Remahoe o Kemakaihi, 1824. Tamehamelu, queen of the Sandwich Islands, departed this life in London, July 8, 1824, aged 22 years.	

As soon as the coffins were placed on the biers, the procession moved towards the church in the following order, through **a double line of native guards.**	
I. Twelve native warriors clad in their beautiful feather war-cloaks and helmets, and each trailing as in mourning a kahile or ensign. II. The marines of the Blonde, their arms reversed. III. The band playing a dirge. IV. Chaplain and surgeon of the Blonde, and two missionaries. V. The funeral cars, each drawn by forty chiefs. VI. Kiaukiauli, brother and successor of Iolani, in Windsor uniform, crape on his arm, &c. with the consul. VII. The princess Naheinaheina, supported by Lord Byron. VIII. The chiefs, male and female, in deep mourning, according to rank, each supporting a British officer. IX. Foreigners, resident agents, masters of vessels, &c. And lastly, 100 seamen from the Blonde, dressed in white, with black handkerchiefs, two and two.	*First.* Twenty men in the native costume of black, some with the addition of rich feather cloaks, each two bearing an immense feathered staff of state, about thirty feet long, and from one to two feet in diameter, some of black, some of crimson, others of green, and others again of yellow feathers. *Second.* The marines of the Blonde, arms reversed. *Third.* The band, playing a dead march. *Fourth.* The gentlemen of the Mission, with the surgeon and chaplain of the frigate; Mr. Bingham, Mr. Bloxam, and myself, walked together, the chaplain, in his full canonical robes; Mr. Bingham and myself in plain dresses, with white scarfs and hat-bands. *Fifth.* The Coffins. These are highly finished and splendid; being covered with rich crimson Genoa velvet, studded with gilt nails, and having the corners of each compartment of the sides, ends, and tops, as well as the massive handles, filled with devices of the cherubim. They were placed on two cars, with black canopies, and each drawn by forty of the inferior male chiefs: it is customary to have all services to a dead chief of high rank performed by chiefs only, in European dresses of deep mourning. *Sixth.* The young king, in his Windsor uniform, with crape on his arm and sword-hilt; and the princess his sister, in full black: the former supported by Mr. Charlton, in his consular character and costume, and the latter by Lord Byron, in full naval uniform.

Having reached the church, which was hung with black on the occasion, the cars were drawn up before the door, and the persons of the procession formed a circle around, while **the chaplain of the Blonde read the funeral service in English, and the American missionary addressed the assembly in their native tongue.** The procession then in the same order marched to the same house, belonging to Karaimoku, where we had been received the day after our arrival: it was now entirely hung with black, and a raised platform, over which a low arch was thrown, at one end, was prepared as the resting-place of the remains of the two sovereigns, whom the old man had loved as his children through life, and whose early death has been most grievous to him. He received their bodies standing by a chair covered with black velvet, placed for him close to

Seventh. The rest of the high chiefs, according to their rank and relationship to the deceased, two abreast, with an officer of the Blonde, in full dress, on each side.

Eighth. The inferior chief women, in European dresses of black.

Ninth. The foreigners, mercantile agents, residents, and sea-captains in port.

And lastly, about one hundred of the men of the Blonde, in a uniform dress of white jackets and trowsers, white hats, and shoes bound with black, and black handkerchiefs and hatbands: the whole making a most respectable, and, for the Sandwich Islands at least, a truly splendid appearance.

The great weight of the coffins, each body having three, one of lead, one of mahogany, and one of oak, rendered it almost impracticable to have them taken into the chapel, which had been hung in black for the occasion: on reaching the door, therefore, the procession, instead of entering the building, fell into a circle around the cars, and **Mr. Bloxam read a part of the burial service of the Episcopal Church, and Mr. Bingham made an address in the native language;** after which the procession again formed, and returned down the street a few rods, to the gate leading to Karaimoku's residence, situated in a cultivated enclosure of some eight or ten acres. Here, as it passed by a circuitous way, it showed to the best advantage.

On reaching the door of the house, the feather-bearers, marines, and band, opened to the right and left, and Mr. Bloxam, Mr. Bingham, and myself, entered the habitation about to be converted into a tomb. I scarce know when I have been more tenderly affected. It the platform prepared for them; and prepared as he was for the reception of their remains, he was extremely agitated, and could not restrain his tears. As soon as the coffins were deposited on the platform, the band accompanied some native singers in a funeral hymn, which the missionaries had written and taught them to sing, to the air of **Pleyel's German Hymn.** We could not help reflecting on the strange combination of circumstances here before us: every thing native-born and ancient in the Isles was passing away: the dead chiefs lay there, hidden in more splendid cerements than their ancestors had ever dreamed of; no bloody sacrifice stained their obsequies, nor was one obscene memorial made to insult the soul as it left its earthly tenement; but instead, there was hope held out of a resurrection to happiness, and the doctrines admitted that had put an end to sacrifice for ever, and pronounced the highest blessing on the highest purity! Where the naked savage only had been seen, the decent clothing of a cultivated people had succeeded, and its adoption, though now occasional, promises permanency at no distant period. Mingled with these willing disciples were the warlike and the noble of the land the most remote on the globe, teaching, by their sympathy, the charities that soften yet dignify human nature. The savage yells of brutal orgies were now silenced; and as the solemn sounds were heard for the first time, uniting the instruments of Europe and the composition of a learned musician, to the simple voice of the savage, and words, not indeed harsh in themselves, framed into verse by the industry and piety of the

was the audience room, in which but a day or two before, Lord Byron and his suite had been presented; but how greatly was its whole appearance altered! A low and well-defined arch had been thrown over its lofty roof, converting it into a long vaulted hall, every part of which, even to the pillars running through the middle, was entirely covered with black. Its only furniture was a large platform at one end, on which was to be deposited all that remained of Kamehamaru and RihoRiho: and the only person in it was Karaimoku, standing to receive the lifeless bodies of those he had ever called and loved as his children. His dress was the same as on Saturday; but his appearance even more venerable and interesting. He received our salutations in agitated silence: the trembling hand, the quivering lip, the hasty and disturbed pressure of his handkerchief to his eyes and forehead, all expressing deep emotion of heart. He permitted himself to be led to a seat, while the coffins, each borne by twelve men of the Blonde, were placed on the platform. Immediately after, the rest of the company entered, and formed numerous groups around the apartment, while, accompanied by the band drawn up on one side, we sung the native funeral anthem to the tune of ***Pleyel's Hymn,*** and Mr. Bingham made a short prayer in the same language. This scene was by far the most striking incident of the day. The sable drapery of the room, the full mourning suits of the majority, interspersed and relieved by the rich and glittering uniforms of the officers, the handsome half-mourning dresses and white plumes of Mrs. Charlton and teachers from a remote nation, came upon the ear, it was impossible not to feel a sensation approaching to awe, as the marvellous and rapid change a few years have produced was called up to the mind.

sister, Madam Boki, and one of the young queens-dowager, who has lately become a bride, with a variety of other becoming and appropriate dresses, the melancholy tones of the instruments, and the solemn truths of the chant in the chamber of the dead, combined to make it a truly interesting and affecting moment.

Thus, under the mournful reverberations of the minute guns, have been received, and peacefully deposited on their native shores, the king and queen of the Sandwich Islands, whose short reign has been marked by epochs no less momentous than the abolition of a cruel system of idolatry; the introduction of the elements of literature, and the arts and usages of civilized life; and, above all, the promulgation of the Gospel, with all the bright hopes and eternal retributions of Christianity! To these dark monarchs of this western archipelago, in the gloomy mansion of their repose, I gave one more tear; one more tear, in remembrance of the better characteristics, which, not unfrequently, shone through their follies and their sins; one more tear for the attachment I had felt, and the prayers I had offered, for them; one more tear for the early disappointment of the hopes I had entertained, that they would return to their rude subjects, from the bright regions of Christendom, only "to point to heaven, and to lead the way." I shudder to think, that, so imperfectly instructed, so partially reclaimed, and, as it is to be feared, so little prepared, they have been launched into eternity, and have stood before their God.

Many details from the two accounts are similar and they are roughly the same length. Three stages shared by both narratives are: that the high chief Karaimoku [Kalanimoku], also known as Mr. Pitt, was not present at the ceremony; that there were twelve stages of the procession, which are described; and that the service included Pleyel's *Hymn*. Callcott includes the order of the people coming on shore from the *Blonde* and the inscriptions on the coffins. Stewart includes more names of the European participants, while Callcott includes more names of the Hawaiian participants. There are some notable differences in vocabulary, such as "dirge" in Callcott versus "death march" in Stewart. Finally, both of them frame the event in terms of the move towards a more Christian and civilized existence, wherein the funeral is a sign of how the islands have improved.

A third, much shorter, printed account of the visit of the *Blonde* can be found in the American missionary Hiram Bingham's *A Residence of Twenty-One Years on the Sandwich Islands*, first published in 1847. Most of his description in this section concerns the activities of the missionaries during the visit. There is some account of a meeting held to discuss changes in the government, in which Bingham played a role. He includes details of the discussion and a list of recommendations given by Lord Byron to the assembled *aliʻi*. The voyage of the *Blonde*, however, plays a background role in Bingham's text, and Bingham appears to have had limited interactions with those on board. Bingham's account of the royal funeral is about a third of the length of the other printed versions, and lacks much of the detail offered by Callcott and Stewart. Interestingly, Bingham spends considerably more time describing the *kahilis* that were carried by the Hawaiian men. His account, beginning with the bodies being taken from the ship, is as follows:

> After the barges, conveying them [the bodies] from the ship, reached the wharf, whence the royal pair had embarked, in 1823, the bodies were placed and borne upon two strong hearses, each drawn by a competent number of men of rank. A procession attended them, first to the door of the church, then to the temporary mausoleum, in the following order: First, twenty men, bearing superb state *kahilis* or feathered standards, twenty to twenty-five feet in height or length, black, green, red, and yellow. About one third of the length of the rod is set with hundreds of little stems or artificial branches radiating in every direction from the rod. These stems, branching into several divisions, have their extremities ingeniously set with tufts of short bright feathers equidistant from the rod, assuming a

> cylindrical form from twenty to twenty-eight inches in diameter. Some of the rods were ornamented with bone and ivory, and multitudinous tortoise-shell rings, and the whole were very imposing. These were followed by the marines of the *Blonde*, in their uniform, with arms reversed, the band playing a plaintive march; the gentlemen of the mission, with the chaplain and surgeon of the *Blonde*, then the honored remains of the late king and queen, followed by the mourners, two abreast, according to rank or relationship. Kauikeaouli, in his uniform, with crape on his arm and sword hilt, and Nahienaena, in full black, the former supported by the British consul, and the latter by Lord Byron; then, each supported by an officer of the *Blonde*, Kaahumanu and Kalakua, Adams and Boki, Hoapili and Namahana. Then followed the chiefs, male and female, of inferior rank, and many others.
>
> The procession moved slowly on between two lines of native soldiers, and crowds of the people who thronged around. The solemn pageant halted at the door of the sanctuary. A few passages from the burial service were read by the chaplain in English, and a short address made in Hawaiian by a missionary. The procession turning its course, and again moving as before, the bodies were conveyed to the thatched house of Kalanimoku. This, but recently, had been the hall of audience, but now, arched and lined with black *kapa*, it was prepared as a temporary repository. Here the old chieftain, Kalanimoku, who called them his children, was waiting to receive them. Halting at the door the marines rested on their arms reversed, and the bodies, being carried in, were placed upon firm platforms. A funeral hymn closed the mournful service. The government soon erected a more permanent but very simple mausoleum of stone, about twenty-four feet square, in which they were finally deposited with funeral solemnity. (Bingham 266)

Bingham then describes the impact of the death and funeral on the expansion of Christianity in the islands. For Bingham, the events were both a way to increase the number of Hawaiians converting to his religion and an opportunity to increase his influence over the *aliʻi*. Stewart is more comfortable placing himself in his narrative. Bingham's framing of the event is largely the same as Stewart's—the death and funeral were opportunities to reflect on the rise of Christianity in the islands. It is a bit odd that Bingham writes "a missionary" made a short address in Hawaiian, when Stewart tells us that it was in fact Bingham himself.

Callcott has a perspective that is markedly different from that of Stewart and Bingham. Stewart and Bingham are very much aligned with the missionary project, and the American missionaries in particular. As a result, Stewart mentions Bingham, the American Protestant missionary, and Rowland Bloxam, the chaplain on board the *Blonde*, by name, and ends his account with a page-long reflection on the spiritual situation of the people of the islands. Callcott, on the other hand, while still concerned with the spread of Christianity, is much more concerned with the move from barbarism to civilization, and the central role that Britain is playing in that process. For Callcott, the British are seen by the Hawaiians as trusted friends who will protect the islands without dominating them. If Callcott's frame is more diplomatic and based on a benevolent global empire, Stewart's and Bingham's frame is more evangelical, filled with prayers and damnation.

James Macrae

James Macrae had a different relationship to the voyage than Callcott, Stewart, or Bingham. He was actually on the voyage. He was a young Scottish-born naturalist employed by the Royal Horticultural Society to collect plants throughout the voyage, to bring plants to the Hawaiian Islands, and to describe the events, places, and people that he observed. Macrae's journal is an unpolished personal account of the voyage, often including information gained by others on the ship, on the islands, or in print. He did not have an important status aboard the *Blonde* and would often complain about how the gentlemen treated him. He did not have access to needed resources and would often complain about the lack of space and equipment to carry out his mission. There is much that Macrae did not know about what was happening during the voyage. Macrae was largely unaware of Hawaiian culture or how the islands had changed since James Cook arrived forty-seven years prior. He did not know that the historian and genealogist David Malo, 32 years old when the *Blonde* arrived, had moved to Lahaina two years earlier to study under the missionary William Richards. John Papa ʻĪʻī, who was also studying with Richards, was 24 at the time the *Blonde* arrived. Both Malo and ʻĪʻī likely interacted with people on the *Blonde*, although this is not confirmed in the narratives. Macrae attempted to include some Hawaiian words in his journal, but he has no sense of the language. He is not a well-informed observer, but he is a trained observer. His journal provides an account of his experiences through the lens of an early 19th-century British botanist trained in providing detailed descriptions.

Macrae was also at the royal funeral, but he had no important role; he was merely someone in the crowd. In his journal, he frequently refers to himself, his opinions, and his interactions with people before, during, and after the event. His perspective, then, could be expected to differ from those of the officers, gentlemen, and missionaries. With Macrae, however, there are two versions to consider. The first is the extract of the journal that was published by William Wilson in 1922, which is being included near the end of this book. The second is a recent transcription of the original journal, which is being published for the first time. Very limited editorial changes have been made to improve the readability of the original journal. In this way, it is possible to consider not only how Macrae's version of the events surrounding the funeral differs from those of Stewart and Callcott, but also to consider how the published version of Macrae's journal differs from the original.

Comparisons of key passages from Macrae's original journal with Wilson's extract can be found in the Conclusion. The focus here is on considering how Macrae's perspective differs from those of Callcott, Stewart, and Bingham.

The passages bolded in the Stewart and Callcott versions are also bolded here.

Wilson's Extract, pages 158 to 160	**Macrae's Journal, pages 50 to 52**
May 11. The coffins were lowered over the ship's side into the launch. Those going on shore to the ceremony got into boats at 1 p.m., the boats forming a line ahead of the launch to tow her ashore. Lord Byron went in his gig on the right, the first lieutenant in the whale boat on the left, with colors in all the boats hoisted half mast. On leaving the ship **26 minute guns were fired,** which was continued by the same number of guns from the forts on shore and a few from an American ship in harbour. The procession of boats towing the bodies had a grand and solemn appearance. On landing, the coffins were placed on two light four-wheeled carts, covered with native black tapa cloth, the head ends of the coffin in view, and drawn side by side by	11. A hazy morning with light wind. At 9am the coffins of the King & Queen were lowered over the ship's side into the launch & shortly after orders was given to those on board to have themselves in readiness by 12 o'clock. At one pm all those going on shore got into the different boats assigned for them, who then formed a line ahead of the launch to tow her with the bodies on shore accompanied by Lord Byron in his gig on the right, & by the first Lieutenant in the whale boat on the left, with colours in all the boats hoisted half mast. The boats on leaving the ship **26 minutes guns was fired,** which was continued by the same number of guns from the forts on shore & a few from an American ship in the harbour. The procession formed by

their late majesties' favorite domestics, preceded by our marines and the band playing slow marches. Next after the wagons came his present majesty, supported on the right by Lord Byron and on the left by Mr. Charlton, the British consul, all in full uniform.	the boats while towing the bodies on shore had a grand& solemn appearance. On landing the bodies they were placed on two lightfour wheeled wagons, where a frame of wood had been placed above the sides to support the coffins arched over, covered with black native tapa cloth, left exposed behind to have the head ends of the coffins in view & drawn beside each other slowly on by their late Majesty's favourite domestics, having before them our marines & the band playing slow marches, & after the wagons joined next to the bodies his present Majesty supported on the right by Lord Byron & on the left by Mr. Charlton the British consul, both in their full uniforms & the King in his new Windsor uniform. After these three followed our Lieutenant & their other gunroom messmates, joined by a number of Queens & High Chiefs with their wives & all draped in black after the European manner, but many of them without shoes or stockings. Next to
After these came the lieutenant and gunroom messmates and the queens in black, but many of the latter without shoes and stockings. I followed with some passengers, and then the lower rank natives and at the end about 100 sailors, four deep, with a midshipman right and left. Thus we advanced **between a line of native soldiers** with rusty arms reversed, naked except for the maro, save a few that had on Russian military jackets and six on each side wore handsome coloured feather tippets. These twelve we concluded must be of higher rank. The five large feather plumes, used on the day of introduction, were carried before the band in a leaning posture, as they do when going into battle.	these again came myself with some more steerage passengers, intermixed with several natives of a more inferior rank, followed by about 100 sailors cleanly dressed forming a procession of four deep with midshipman in full uniform on their right & left. In this manner we kept slowly advancing **between a line of native soldiers** standing at short distances from each other, with their rusty firearms reversed & most of themselves naked with only the maro on. Some few of them we saw had on Russian military jackets, that gave them very ludicrous appearance, from the rest of their body being exposed naked, but that which the jacket covered. One singular from the rest attracted most of our notice from his having on an old black hat & a white

The procession halted at the missionaries' chapel, where our chaplain read a prayer in English, and then one of the missionaries did the same in the native language. Resuming our way, the procession reached Mr. Pitt's hut, between two and three o'clock. The marines formed a line in front outside the hut. The weighty coffins were taken from the wagons by the sailors and deposited on a platform inside the hut. Here the queens of the deceased king, joined by three of his father's still alive, and who are of higher rank, stood at the head of the coffins, and were joined by the wives and daughters of the chiefs forming a line on one side of the hut. Pitt, Boki and other chiefs of consequence were admitted, together with the principal officers of the *Blonde* and a few Americans. No inferior class of natives were admitted nor had been allowed to join in the procession. **Then the chaplain gave out an anthem. It was accompanied by the band. A prayer followed, then the same performance by a missionary in the native tongue.** Pitt, who had never joined the procession, remained at home in his arm chair, as usual dressed in black.	dirty shirt half way down his thigh, leaving the rest of his legs bare. Both of these articles are seldom seen worn by the inferior class of natives. After we had passed the man in the shirt then we came to 12 (six on each side) in handsom-coloured feather tippets, which we concluded to be of a higher rank. Before the band was carried five large beautiful feather plumes mentioned to have [been] seen behind the King on the day of being with others introduced to Pitt. These were carried in a leaning posture as is customary with them when going in battle, which from their variety of colours added much to the solemnity of the scene on the occasion. On our way we halted outside the missionaries' chapel, where in front they have a bell hung on an old frame of wood. **Here our Chaplain read over a prayer in English and after him one of the missionaries another in the native language,** which when done we again resumed in procession as before, till arriving at Mr. Pitt's hut between two & three o'clock, when the marines formed outside a line in front. The weighty coffins were now taken from both the wagons by the sailors & deposited inside Pitt's hut at one end where a low platform of wood had been previously erected on purpose covered over with fine mats of various striped colours & hung round above with black Jap cloth. After the bodies had been safely deposited under the same roof with Pitt himself. The Queens of the deceased joined with three of his father's still alive & who are of a higher rank stood at the head of them in forming a line on our side the hut joined by the wives & daughters of the chiefs, the same as on the 7th & on the opposite side, Pitt,

(Note: *Pleyel's Hymn* is not mentioned in Wilson's version.) The whole ceremony was conducted with solemnity from the time of leaving the *Blonde* to the finish. One mistake, however, had been made. No invitation had been given to the few respectable Americans resident in the place; why is best known to those who had the conducting of the funeral, for they were left to form a small body by themselves on one side, away from the procession. It could not have been the wish of the chiefs, for they were seen often on their way to leave their places and join the Americans, especially Boki.	Boki & other chiefs of consequence with the principal officers of the Blonde & a few Americans, no inferior class of natives being admitted, or before allowed to join in procession. We had not long entered the hut when the chaplain gave out an anthem which was sung accompanied by the band, then he afterwards read a prayer, & the missionaries in a similar manner went through the same form in the native language. Pitt who all along never left his hut to join the procession, remained at home quite composed in his arm chair draped in his usual black gown. The whole of the funeral procession throughout commencing with the disembarkation of the bodies from the ship till deposited, was conducted with solemnity in a style far superior to what most on board before going on shore had even expected could be done. But notwithstanding there was one fault allowed to occur to any observer like myself free from partiality, that in not giving a general invitation to the respectable few Americans resident in the place or even afterwards when they attended at the landing of the bodies from the ship, no invitation was given them to join us in procession, but suffered to form unasked a small body of themselves at some distance apart on one side all the way. What might be the reasons assigned for treating them with such distant marks of disrespect is best known to those who had the conducting of the funeral. It could not be from any wish on the part of the chiefs, for they were often seen on the way to leave their places, particularly Boki, & join the Americans with whom they appeared to be on the most friendly terms. When the funeral ceremonies were over all persons belonging to the ship were ordered to

On the conclusion of the ceremony, all belonging to the ship were ordered back to their boats in marching order, with the band in front playing lively tunes. The natives broke through their soldiers' guard and clustered around us, evincing great astonishment at the great drum, and accompanying us to the harbour, where **the ship's company were served with grog, and the officers had similar refreshment at Lord Byron's house.** By five p.m. all were safely back on board. Mr. Dampier, the draftsman, had been stationed at an early hour where he could best make a drawing of the funeral procession, and is said to have already begun taking portraits of some of the queens with whom he has taken up his residence among them at their homes.	return to their respective boats in marching order with the band in front playing lively airs, which brought nearly all the natives in the place from their homes eagerly running towards us like mad people on hearing the sound of the great drum that so much astonished them, they regardless of their own soldiers broke instantly on both sides through their lines & came in amongst us all the wall till we reached the harbour to embark, which afforded to many, particularly the sailors' good amusement. **Before embarking the ships' company had a glass of grog each & the officers a similar refreshment at His Lordship house.** At 5 o'clock we all arrived without any accident on board, further than a few of the marines managed to get a drop too much. Mr. Dampier the draftsman had been at an early hour stationed in some convenient place, where he could with best advantage take a drawing of the funeral procession from the disembarkation to the place of internment. This gentleman has already I am told commenced taking portraits of some of the Queens, with whom he has taken up his residence among them at their homes, some say for the purpose of getting a more correct painting. Thermometer at 6am 73 at 12 76 & at 6pm 73.

The lengths and variations in content of the five different versions of the event are notable. Callcott's account has the greatest number of words (1,315), with the inscriptions on the coffin accounting for about 100 and her moral and religious reflections on the situation accounting for another 500. The rest of her description amounts to roughly 700 words. Stewart's account is 1,212 words, with the final paragraph of 200 words devoted to religious reflections, leaving roughly 1,000 words to describe the event. Bingham's account is a mere 444 words and has few details except the description of the *kahilis*, which is not found in the other versions. Macrae's original journal is 1,283 words, while Wilson's extract of the journal is only 734. Neither the journal nor the extract step back to offer larger-scale reflections on the situation: their focus is on describing, and sometimes judging, the event itself. They mention drinking grog afterwards, a detail left out of the other accounts. With the description of the funeral in Macrae's journal and Wilson's extract, we are offered observations from someone in the background, commenting on what people were wearing, why people were not invited, and how people behaved. Macrae is judgmental. He points out problems with the ceremony. He wonders why the Americans were not invited. He notices details such as the bell and the wooden frame outside the church, and refers to Bingham as "one of the missionaries" instead of by name. He describes the event from the perspective of a minor participant and does not place the funeral in a larger story of civilization or conversion, although he is likely in favor of both. Finally, his journal includes descriptions of the logistics of attending the funeral and of returning to the ship afterwards.

Macrae's own journal is complex given the origins of the descriptions that are included. Macrae did not write about the islands as an isolated observer. Some of the text can be traced back to the people who inform what he writes. Macrae is writing his journal, but even the sources of the single manuscript are varied—other printed texts, conversations, and his instructions from his employers all have an impact. He has informants who live on the islands, including John Young, a former advisor to Kamehameha I, "a Bengal black, who informed us, [had] been wrecked in an Indiaman on the coast some years ago"; (page 57) an "Irishman who had the management of the guns of both forts"; (page 49) Americans and Europeans in various situations on the islands, and native Hawaiians, although these were less common because Macrae did not know their language. Macrae also spoke regularly with Lord Byron and others on the ship, who provided him with additional information from their own forays around the islands.

Ultimately, Macrae produced a journal that offered an account of the Hawaiian Islands that is insightful because of his immediacy and expertise, but also questionable because of his limited understanding of Hawaiian culture and of how what he observed connected to what was happening in the islands and around the world. As part of a larger set of texts, written primarily by Europeans and Americans, these writings offer some useful descriptions of the Hawaiian Islands. The texts can also help us understand how Europeans and Americans thought about themselves, the Hawaiians, and the world—thoughts that are themselves woven into the history of the islands.

PART I.

JAMES MACRAE'S ORIGINAL JOURNAL, 1825

INSTRUCTIONS TO JAMES MACRAE FROM THE HORTICULTURAL SOCIETY OF LONDON

by Joseph Sabine

Instructions to Mr James Macrae employed by the Horticultural Society of London on a mission in His Majesty's Ship *Blonde* commanded by Captain Lord Byron to the Sandwich Islands in the Pacific Ocean and other places in 1824 and the following year.

1st. You will embark on board the *Blonde*, and from the time of your embarkation, till you return to England you will consider yourself under the direction of Lord Byron; you will obey his Instructions in the performance of your employment, and you will consult him and follow his advice, in all points, whether in the execution of the objects of your mission, or in any matters relating to yourself personally.

2nd. The object of your mission, generally, is to collect plants and seeds of all descriptions, whether wild or in a cultivated state, for the use of the Society, attending particularly to those which are likely, in any way to be valuable, either as esculents, fruits, or for other useful purposes; and next, such as will be ornamental, of whatever description they may be.

3rd. You will be particularly careful, on ship-board to conform yourself to every regulation made, or orders given by the Captain. You will be informed when and in what manner you are to go on shore to follow your pursuits, and on every excursion on land, you are especially enjoined to be most punctual to any appointment made for your return on board.

4th. If you are in want of pecuniary supplies, either on your own account, or on that of the Society, and if such appear requisite to Lord Byron, his drafts on the Treasurer of the Society in London, for money for such purposes, will be honored; you will give the earliest intelligence by letter to me of such drafts, and state the especial purpose to which the money is applied.

5th. The preceding instruction respecting money is not limited to amount, it is presumed however that any supply you may want will be small, and it is to be understood that you will not embark in any expensive undertakings of collection, beyond those which are pointed out or noticed in these instructions.

6th. Many of the esculents you will meet with are probably not known as such in England; you will omit no opportunity of enquiring after such, and of looking for them in the gardens of the places you visit; you will collect all you can of them, making notes of the method of cultivating them, and the manner in which they are used. The fruits you will see will be, of course all natives of hot climates but these the Society, are especially desirous to obtain; you will therefore endeavour to procure seeds or plants of them, you will [be] particular in your descriptions of them, as well as in noticing and recording every circumstance relative to their habitats in a wild state, and to their cultivation or necessary treatment where in gardens.

7th. After fruits and esculent vegetables, you will search, enquire and procure, such plants as may be useful in medicine, or for dyes. In the ornamental class you will particularly see to the acquisition of bulbous plants, and especially of air plants. To creepers, shrubs and perennial herbaceous plants, you [will] also generally direct your attention.

8th. To the packing of plants, bulbs and seeds it is presumed you are sufficiently skilled not to require instruction. The methods therefore of preparing your packages is left to your own discretion.

9th. Specimens of the fruits of the countries you visit, if particularly curious, may be preserved and brought home, immersed in spirits, in bottles well closed up.

10th. It is desirable that you should collect specimens of capsules and seeds of plants, to be preserved for exhibition in the Society's apartments; these should therefore be selected as perfectly formed as possible and packed dry in boxes.

11th. You will of course gather and dry specimens of the plants of every country you visit. You are furnished with all the requisites for forming an herbarium. Your collection must not be confined to single specimens; it is desirable that you should bring or send home several of each species, that you meet with. You will be particularly careful to place a note with the several specimens, of the time when, and place and situation, where they were gathered. As it is of little importance whither you are able with certainty to affix a name to each correctly, it will not be necessary to occupy your time for such purpose, when you can otherwise employ yourself more usefully.

12th. You will be particularly careful to affix some mark to each packet of seeds, and to each plant or bulb, which you transmit home, referring it to the specimen corresponding to each in your herbarium; and you will be careful, that such mark shall be so made, as to be permanent until it reaches England.

13th. You are furnished with letters of introduction and packets to deliver to our correspondents at Madeira, Rio Janeiro and Valparaiso, if you touch at other places where there are correspondents of the Society, you will find their name in the list of the Society which you take with you. The production of these instructions to such persons will no doubt be a sufficient introduction to them, and will obtain for you their assistance as far as may be in their power in the furtherance of your pursuits. You have with you also copies of the garden reports of 1823 and 1824, to shew to such persons as you may think expedient.

14th. If you find that no correspondent of the Society lives at any of the places you land at, you will enquire for, and endeavour to become acquainted with persons who are likely to be serviceable to the Society, by their correspondence and communications, and having ascertained whether they are willing to take on themselves the trouble of such correspondence, you will report the particulars of their names, residence, and address to me. If at any place you meet with a person likely to become [a] useful corresponding Member of the Society even though it have one previously resident there, you will engage his services and report to me in like manner respecting him.

15th. You are furnished with a copy of the instructions for transmitting plants &c from tropical countries which have been printed for the use of the Society's correspondents and agents. It is desirable that any additional observations you may make on this subject, should be carefully and separately noted, in order that such as are likely to be useful may be incorporated or added when another [edition] of those articles shall be printed.

16th. You will keep a fair journal of all your proceedings and observations, and make a fair copy of the same from time to time; and you will send home the parts of this copy, when you are able to do so, addressed to me, at the House of the Horticultural Society in Regent Street. Your journal should be made as ample as possible; it must contain the details of each day's proceedings, as well as notices and accounts of all that you see. Whenever you send home a copy of part of your Journal, you will at the same time send a letter to me informing me to whose care the same is entrusted.

17th. You will be particularly careful to write to me, by every opportunity which occurs, and address your letters to me at the House of the Society in Regent Street, London. You will number the letters which you send progressively, keeping copies of them all; and in each letter you will mention at what time and from what place, your former letter was dispatched and enclose a copy of such former letter.

18th. When you transmit home letters and other papers or parcels from any place be careful to send your letter separate from the papers or parcels. Captains of vessels are compelled to forward all packets of papers having the appearance of letters by the post, from the first port they arrive at in England. The postage of such packets is often very large and if they do not contain letters is usually returned but if letters are also in the parcel no allowance is made from any port.

19th. Whenever you are not immediately engaged in your Horticultural and Botanical pursuits, you will do your utmost to make collections in other branches of Natural History; the formation of which from the apparatus and materials you take out with you; it is presumed you will be well able to effect.

20th. From the interest taken by Lord Byron in your appointment; and from the information communicated to [me] respecting the disposition and character of several of the officers in the frigate, I have no doubt that you will find some of them desirous of instruction in the scientific objects of your employment. You will, therefore, at all times be ready to impart whatever knowledge you are able to such persons and to contribute as far as is compatable with your duty to the Society, your assistance to their pursuits; you will also shew liberality in giving specimens that you can share; you will find advantage to yourself in such conduct; for should articles of rarity fall into their hands such will be more readily imparted to you when it is known that you are willing to give from your own collections such as it is not incumbent on you to retain for the Society.

21st. It is requisite however that subject to the discretion above given you should fully understand that everything collected by you during your voyage is exclusively the property of the Horticultural Society and that you have neither permission or power to give or send to any individual whatever in England or elsewhere any article which shall have been collected by you and the whole of the Journals and notes made during your voyage are in like manner to remain at the absolute disposal of the Council of the Society.

22nd. You will avail yourself of every safe conveyance that offers to send home whatever you shall have collected carefully packed, sending at the same time advice thereof. The packages must be addressed to "The Horticultural Society of London to the care of David Maclean Esq. at the Custom House London.["] If sent by any of His Majesty's Ships they need not be consigned to Mr. [Maclean] but only directed to the House of the Society.

23rd. In case you send packages to the port of Liverpool they must be directed to the care of "Messrs Morrall & Woodward of Liverpool for the Horticultural Society of London." If they are sent to any other port direct them to me at the House of the Society, London, and request the Captain to write to me on his coming to the port to inform me of the arrival of the package. Send a letter of advice yourself in all cases.

24th. You will take as much care as possible of the different books, implements & materials furnished for your use and with the exception of such as are applied to their destined purposes during the voyage, you will preserve them in good order to be restored to the Society on your return.

25th. You are supplied with a small collection of seeds of esculent vegetables for the purpose of distribution or of presents to persons at any places where you may touch. Should Lord Byron be desirous at any time of disposing of any portion of these for similar purposes you will give him what he wishes for.

26th. You also have with you a well selected collection of plants chiefly fruit trees destined for the use of the inhabitants of the Sandwich Islands; in the disposal of these you will be entirely guided by the opinions of Lord Byron; you will direct their planting and attend to them during your stay, and at your departure it will probably be desirable to place them under the charge of Capt. Charlton the British resident at the Islands. The boxes in which these plants are packed will be used to bring back living plants and will serve as models by which to make others.

27th. You have a packet to deliver to Captain Charlton. [He] will possibly not have arrived when you get to the Islands, but you will probably see him before you leave them. He is known to be ready to forward the objects of the Society and you will give him instructions and make arrangements with him for the transmission of any collection which may be made after your departure.

28th. It is desirable that you should communicate with Mr. David Douglas who is on his voyage to the mouth of the Columbia to collect for the Society in the north western part of North America. The letter sent by you for him must be forwarded as soon as possible to the mouth of the Columbia from whence it will reach him.

29th. When you are at Valparaiso, you will get all the information possible and make observations on the practicability and expediency of sending a botanical collector to Chile. The plants of that country are known to have much merit both as regards beauty and utility and such of them as will suit the British climate are particularly desirable. You will especially inquire for and send home plants and seeds of the Chili and Araucaria; the later should be packed in boxes of mould.

30th. It is well known how much seeds are injured by the double passage across the line. You are therefore directed to make endeavours to send portions of your seeds collected north of the equator from some port in the Pacific from whence the[y] may be forwarded direct to Europe; but do not trust the despatch of others but duplicates because their safe arrival will be very uncertain.

31st. Should your return round Cape Horn be in the summer season of the Northern Hemisphere, I have no doubt of your success in bringing home a very fine and valuable collection of living plants; little hope of their surviving can be entertained should your passage round South America be in the winter season.

32nd. Respecting the care of your living plants on shipboard you are at all times when you experience any difficulty to apply personally to Lord Byron for his instructions and orders. There is no doubt you will receive from him every assistance and support which he can afford consistent with the due order and regulation of his ship.

33rd. You will probably touch at the Falkland Islands on your return, a short stay will enable you to make a perfect collection of the plants of these places which will be particularly interesting and valuable.

34th. On your return you will write to me from the first port of the British Islands which you make; you will remain with and keep charge of the different collections you have with you, until they are safely deposited on shore or transmitted by land to London; and then you will as expeditiously as possible, come yourself to the Society's House to report the results of your voyage and assist in the arrangement of whatever you may bring home with you.

Regent Street London
September 16th 1824

BOXES OF PLANTS FOR THE SANDWICH ISLANDS

No. 1

Annona cherimola
Annona squamosa
Anona Sp.
Annona muricata
Psidium pomiferum
Psidium pomiferum Sierra Leone
Psidium chinense
Psidium Sp. from Maranhão
Barbadoes Cherry
Eugenia aquea
Eugenia sambos
Cicca disticha
Spondias mombin
Dimocarpus longan
Cookia punctata
Sarcocephalus esculentus
Iatsopha Manihot
China roses 2 Red 3 Pink
1 Sweet briar Rose

No. 2

1 Mellish's Favourite Peach
2 Pourfree Lative Peach
3 Grosse Mignonne Peach two plants
4 Grimwood's Royal George

5 Violette Hative Nectarine
6 White Hative Nectarine
7 Elruge Hative Nectarine
8 Tinta Grape
9 Burgundy Grape
10 Grove-end Sweetwater Grape
11 Mignonne White Cluster Grape

No. 3

12 Highflyer Walnut
13 Marseilles Fig
14 Small blue Fig
15 White Perdrigon Plum
16 Green Gage Plum
17 Reine Claude Violette Plum
18 Early Orleans Plum
19, 20 } Bigarreau Cherry
21 Verdelho Grape
22 White Frontignac Grape
23 Poonah Grape
24 Royal Muscadine Grape

No. 4

25 Boston Russetting Apple
26 White Suneating Apple
27 Early Red Apple
28 Early Harvest Apple
29 Sanguine d'Italie Pear
30 Louise Bonne Pear
31 St. Michel Doré Pear
32 Duchesse d'Angoulême Pear
33 Esperione Grape
34 Black Hamburgh Grape
35 Grove-end Sweetwater Grape
36 Verdelho Grape

LIST OF VEGETABLE SEEDS, TAKEN OUT BY MR. JAMES MACRAE

Blood Red Onion
James' Keeping Onion
Long Orange Carrot
Early Horn Carrot
Early Dutch Turnip
Early Stone Turnip
Scarlet Short top Radish
Red Turnip Radish
Brown Dutch Lettuce
Brown Coss Lettuce
Red Solid Celery
Curled Cress
White Mustard
Curled Parsley
Early Dwarf Cabbage
Early Battersea Cabbage
Long Prickly Cucumber
Short Prickly Cucumber
&
Seven Packets of Rice

JOURNAL OF JAMES MACRAE WHILE AT THE SANDWICH ISLANDS

EDITOR'S NOTE: The transcript of the journal begins May 1, 1825 (leaf 255 of the manuscript). Macrae had boarded the *Blonde* at Woolwich on September 8, 1824 and by this point was eight months into the voyage.

---

1 May. A fine morning with moderate breezes from E.S.E. Church service performed at the usual hour. At 8pm moderate & fine. Thermometer at 6am 74 at 12 76.5 & at 6pm 74.

2. A fine morning with moderate wind from E. by South. 9:30 the cook's mate punished with three dozen lashes for beating the Captain's steward. At 8pm moderate breezes & cloudy weather. Thermometer 73.5 at 12 75.5 & at 6pm 74.

3. A fine morning with moderate wind from East. At 7am made the South east point of Owhyee & at 12 got close in with the land, which afforded us a pleasant contrast compared with the Galapagos lately left. Here green verdure was seen beautiful interspersed with trees & in places along the coast, rows of coco-nuts with leaning bushy tops. The country in this part rose inland with a gentle ascent & did not appear to us much intersected by deep ravines. By 2 o'clock pm we hove to, off Heddo Bay & lowered two of the ship's boats, one for the surveyor & his assistant, & the other for the master to go in shore to try sounding & examine the anchorage to ascertain whither we could return here to refit the vessel after depositing the bodies of the king & queen at the seat of government of Woahoo. While the boats were away two ragged white men came on board & reported that the ship was standing on not far from a reef which lay most way across the mouth of the bay. All was instantly bustle,

& noise to put the ship about to stand out to sea farther off from the land. We learned from these men that news had been brought them for upwards of a month by an English whaler from the Port of London of the King & Queen's death & also that the natives of the island of Attoi to leeward had rebelled two months since against the Regent in consequence of the chief of the island having died since the king went to England & arrived from America some months past of the son & heir to the deceased chief who claimed the rights of his father, which before he was defeated by the superior force of the Regent who commanded in person occasioned the loss of the lives of upwards of 2000 natives besides two Europeans & yet we was assured this same person after being taken is at present suffered to go at large in town of Hana-rura at Woahoo. We had not been long off the Bay of Heddo when several of the natives visited us from the shore paddling off in their canoes, bringing with them some provisions, such as sugar canes, tarro (*Aram* spec.) fish & hogs. The sugar canes were generally of the purple & striped kind & and of a large size. The dress of the men consisted of the maro a piece of various dyed cloth of their own manufacture 8 or 10 inches broad & about 3 yards long which they pass between the legs & tie round the waist. The dress of the women consisted generally of a larger piece [of] cloth of similar manufacture tied carelessly around the waist forming a short petticoat reaching half way down the thigh. This was the common dress of all who came off to the ship, excepting the head chief & one of his wives who is here commonly honoured with the title of Governor, both were of more than ordinary size from overgrown corpulencey. The governor's dress consisted of a dirty cotton shirt, a pair of blue China nankin trousers & an old straw hat. That of his wife the common petticoat used by the other females with the addition of having extra to them a piece of native made cloth tied round her neck hanging loosely over her body open in front like a cloak. Both these two on coming on board touched noses after the manner of salutation of this country with the most of those we brought out from England & immediately after went with them below to their cabins where they all soon commenced a disagreeable howling noise in lamentation of their late King & Queen. The canoes of the natives [that] came alongside the ship had all outriggers the same as described by Capt. Cook & others, that which the Governor came on board in was upwards of 40 feet long & very neatly made. At 5 o'clock the boats returned bringing information that the anchorage was good & the entrance to it sufficiently large for ships of the line. We now made sail for the island of Mowee where His Lordship was informed Queen [Kaumanna] at present resided, to take her on board with any other people of consequence on that island who might wish to be present at the funeral, like the Governor & his wife with several others come already on board from Heddo. The neighbourhood of Heddo Bay to the eastward is thickly covered with wood from the sea coast every where back in the interior as far as we could see the country free from fog & to the westward much intersected with deep ravines ending by the sea in abrupt cliffs down which fall several beautiful cascades of water. The native huts appeared to be most numerous round the edge of the Bay pleasantly situated under thick groves of coco-nut & bread fruit trees. Among the numerous trees that ornament this delightful spot are members of the candlenut (*Aleurites*) so easily distinguished at so long a distance from their white shining leaves, grow everywhere on the coast. At dusk fresh breezes from East & fine weather. Thermometer at 6am 74 at 12 76 & at 6pm 76.

4. A fine morning with moderate breezes from north & cloudy. At 7am saw the summit of Mouna Kaah towering far above the clouds in places covered with snow, we saw also Mouna Roa appearing not far off from the other but without snow. The grand appearance in which both these mountains showed themselves from on board the ship upwards of a hundred miles out at sea would repay the travellers for coming on purpose to see them. We had scarcely satisfied ourselves with looking on the mountains of Owhyee when the high land of Mowee appeared in a timeless way a-head, but no where covered with snow, which in the course of an hour or two, we was under running along the shore, where we noticed the same as to the westward of Heddo Bay that this side of the island was intersected with numerous deep ravines down which fell several cascades of water at no great distance from each other upon the same stream which was admired by all persons on board, most of whom anxiously wished to be on shore that they might satisfy their thirst from such beautiful water, being on board for the last month on short allowance of this article & often of late served out not drinkable. The country on this side of the island of Mowee rose to the peaks of the mountains with much more ascent than any where noticed at Owhyee but like that island well wooded & along the coast covered with green verdure, yet little cultivated although the native huts appeared everywhere more numerous at no great distance from each other generally upon the ridges near the sea coast for the convenience of fishing. At 4pm saw a Sandwich island schooner a-head running close in shore for the town of Lahaina, which by six we got a breast off & made signal for a pilot when several canoe full of natives of both sexes came along side

the ship, bringing us some fruit & provisions similar to that when off Heddo Bay. We here on getting near enough to the shore lowered a boat landed all the Sandwich people brought from England, who were dressed in new suits of mourning, which formed [a] rare contrast beside their naked countrymen. When they landed on the beach they were received by some thousands of the natives making very disagreeable loud howling noise, ceasing at intervals for a few minutes, then commencing again with the same loud noise as before, which they kept up in this manner the most of the night. About 7 a little after dusk we came to anchor in 11 fathoms on a coral bottom not far from a reef with a high surf upon it. During the night the native schooner was sent to Woahoo to inform Mr. Pitt & Mr. Charlton of our arrival, the latter we had already learned had got to that island about a week since which we was happy to hear. Thermometer at 6am 73 at 12 75 & at 6pm 76.5.

5. A fine morning. At day-light lowered all the ship's boats for the purpose of getting a fresh supply of water. At 9 got permission from the first Lieutenant to go on shore after promising not to go by any means out of the ship's sight; being uncertain but we might get under weigh immediately. On my going on shore I met with three poor ragged white men at the watering place who very willingly offered their services to accompany me through the native huts & plantations being to a stranger in places rather difficult to find out the shortest way from the numerous tarro ponds almost every where among the huts. The tarro I noticed was cultivated to a greater extent than any other esculents, the ponds of water in which they cultivate in are commonly square & of different sizes, about four deep from the level surface, having generally from two to two & a half feet at bottom covered with water which is constantly supplied from a neighbouring stream by the means of a small canal emptying itself one to another until at last it reaches the sea shore. These are converted into fish & duck ponds as well as for the growth of tarro, which they in general keep free from rushes & other weeds & planted with surprising regularity not far behind another more civilized race of people. The singular method which they have adopted to growing this farinaceous esculent always in water has originated beyond doubt at an early period with themselves perhaps from their extreme fondness to that element which they so often frequent during the day in a manner similar to the feather aquatic tribe. When they have occasion to weed these ponds they are generally up in water & mud above the knees which to any other race of people not so much accustomed to water would be found inconvenient. They as far as I could learn have no particular

View of Part of Lahaina in Maui. From Maria Callcott's *Voyage of H. M. S.* Blonde *to the Sandwich Islands, in the years 1824-1825. Captain the Right Hon. Lord Byron, commander.* "Engraved from a Drawing made on the spot by Robert Dampier."

season for planting or of taking up the tarro, but generally go on using the largest as they occasionally want them for food from the ponds that are most forward till they become nearly done, when they then turn the water off and drain them dry with whatever fish they may contain [put] into the nearest pond leaving them exposed for a few days to the sun for the mud to harden, then round or square clumps are thrown up about two feet high & seven or eight in circumference at short distances from each other always in straight lines either across or lengthwise in the ponds in these clumps are planted with the crowns of the tarro from eight to twelve in number alternately about a foot apart. The crowns which they choose for planting are the top parts of the tarro cut across about half an inch thick adhering to the leaves which they shorten to the length of eight or ten inches. Round the ponds by the edge of the water is often planted what they term the tea tree (*Dracaena terminalis*) the root of which they cook under ground as is customary with them in every thing else which requires cooking before it is eat, this they afterwards pound & then put into vessels full of water for a few days to ferment, when it makes a good substitute for beer of a very intoxicating quality. A piece of the root which was brought me cooked for to taste was equally sweet as the sugar cane & in my opinion contained an equal quantity of saccharine matter, which if the common process generally adopted for the making of sugar was here once to be introduced it might be turned to the same advantage as the sugar cane for manufacturing sugar, but whilst there is no emulation on part of the natives for commerce or cultivating more than is really necessary for to supply their present wants, the important value that might likely be found out from this plant with so little trouble and expense will like the present remain yet for a length of time imperfectly known. I had not been long on shore when one of the whites met with at the watering place took me to his garden which was small & surrounded his little hut, here I noticed *Cytisus cajan*, or the common Pigeon Pea of the West Indies, which he said was given him last year by the Captain of *American Whales* for coffee with some seeds of the lima bean (*Phaseolus lunatus*) both of which were now in bearing for some time, but being yet unacquainted with the way of using them for food he always felt afraid to touch them for that purpose for fear of doing hurt. These with a few light red coloured cabbages, two sorts of sweet potatoes (the red & white), water melons, pumpkins, a few patches of sugar canes, two or three ponds of tarro & some bitter gourds which are used by the natives for various purposes as calabashes, was all with a hog & a few fowls he had for the support of himself & family. His wife is a native woman by whom he had three children, these he has often applied to the missionaries to baptize without success being by them considered born out of wedlock. In travelling beyond the native huts & plantations to the common pasture land of the island I observed but few plants, some of which were the following: *Cleome*, *Argemone*,[1] 2 sorts of *Malvas*, 3 of *Sidas*, two of which the natives use for wreaths & necklaces by stringing the flowers on a thread made from the bark of the tapa. When in the culturated grounds of the natives I noticed they had no plantains, & only three sorts of bananas, one of these was much shorter than the others & different to any that I have ever seen before. The red banana common in Otahite has not yet reached the Sandwich Islands, but will no doubt be now soon introduced by Mr. Charlton who is to visit that island once every year. The town of Lahaina is composed of a number of low thatched huts along the sea shore about a mile in length and in places nearly half the distance in breadth situated on a level flat at the foot of the high mountains which rise abruptly central in the island. At the westward of the town there is a small grove of coco-nut trees & more towards the town on the sea beach is a mud battery in bad repair, mounted with five small cannon in the same neglected state, within its walls is a small mud hut white washed outside, where I was informed had been buried lately one of Tamehamaah's Queens. Near this fort the missionaries have a small thatched chapel, with other necessary dwelling houses and garden grounds for their own use. There I was shown the only grape vine upon the island which was yet but young & never had produced fruit. Close to the beach nearly in front of the town stands a small brick house of two low stories white washed outside, built sixteen years ago by Tamahamaah for his favourite Queen Kaumanna, which she never inhabited choosing rather to live in a thatched hut after the manner of the country close beside the other. The town has no regular streets, being all cultivated & rather difficult to get from one end to the other on account of the numerous tarro ponds. This pleasant little place looked every where similar to a well cultivated garden, divided into numerous different allotments by mud walls enclosing each family hut & garden, planted with more or less the same sort of esculents & other things according to the fancy of the different possessors. Sugar canes grow with little trouble on the narrow ridges between the tarro ponds, where they have also at times cotton, tobacco, & cabbages. The tapa tree (*Broussetta*) which they manufacture from the bark of the cloth they generally wear, occu-

1. The natives use the [juice?] of this plant which is of a yellow [colour] for venereal complaints, it grows everywhere on the sea coast. [note in manuscript]

pies with in any of them a large proportion of their ground neatly planted out in rows kept free from weeds. The number of natives living at Lahaina I was informed exceeded 6000, which had some years before been far above that number, but since then hundreds had died during a short time from some unknown fever that had been introduced among them. At 4pm the ship's boats having finished taking on board what water, live stock & provisions we at present wanted, orders were given for all persons on shore belonging to the ship to return on board without delay, when at six we weighed anchor & made sail for the island of Woahoo with a strong breeze from the northward, having on board about forty natives among whom we had the mother of the deceased Queen & her daughter an interesting girl about ten years of age, sister to the present youth proclaimed king. This young princess was dressed in deep mourning brought her by Madam Boki from England with every thing else in the English style fit to be worn by a lady, which although before unaccustomed to this sort of dress, she appeared to be much taken up with it. Several of the chiefs who came on board were also clothed in the European fashion, while others of an inferior rank had only merely a shirt of maro without a hat. The rest of the poorer tribe of natives seen here on shore were draped after the manner of those come off to the ships at Heddo Bay. Thermometer at 6 72 at 12 76 & at 6pm 74.

6. A fine morning. At day-light made the land of the island of Woahoo and at 7am got under Diamond Hill, when at once opened to our view the Bay of Wytitte pleasantly covered with thick groves of coco-nut trees where lay underneath many of the natives' huts, behind which appeared the woody mountains rising in places abruptly intersected by numerous deep ravines, having on each side high peaked ridges every where covered with verdure. At 8 we was off the Harbour of Hana-rura & made signal for a pilot, when shortly after Captain Charlton came on board to conduct us to a safe anchorage outside a coral reef which runs across the mouth of the harbour nearly a mile from the shore, having only in one part opposite the town a narrow passage over a bar from three to four fathoms water, which was thought too shallow for our ship going over to a more convenient place in the inner harbour. At 10 on coming to anchor in 20 fathoms we fired a salute of 15 guns, which was to our surprise returned with the same number from both the forts on shore, where we noticed was hoisted their national flag, having the union of England in the upper corner. In the harbour lay at anchor three American merchant ships, besides several smaller vessels, two [of] whom were native

brigs mounted with 8 or 10 small guns. To the westward of the harbour lay wrecked on the reef with only her hull remaining a whale ship of nearly 400 tons from the Port of London wrecked two weeks ago, during the night while standing off & on waiting for the Captain who was on shore enjoying himself too freely with his friends before going on board his ship. The crew with some difficulty saved themselves, & her cargo which consisted of nearly 800 barrels of sperm oil has since been nearly all saved by the assistance of the natives, as also most of her spars, rigging, & sails &c. During the day no person was permitted to go on shore, but the natives & his Lordship accompanied by the chaplain who upon their landing were met by crowds of the natives on the beach making the same disagreeable howling noise as before when at the island of Mowee. Since the time we made the islands & the natives coming on board we have been at times highly amused with their simple manners giving to different parts of the ship eagerly examining what they saw, but at other times we felt ourselves not a little disgusted when we saw them sit down to their meals eating raw fish with the gills & entrails & their fingers covered with blood. They appeared to have no particular stated time for their meals but eat when they felt inclined, sometimes so late as 12 o'clock at night & at others by day-light in the morning. Their diet consisted of raw and dried fish & the tarro. The latter they generally pound with a mixture of cold water on a thick board with a stone to nearly the same consistency with starch prepared by us for dinner, then afterwards is put into calabashes till it is wanted to be eaten; this they term poi which when they can procure fish to eat with it, is their constant & favourite food. When a family have their meals they sit on the ground in [the] form of a ring with their calabashes of poi in the center, where they in turns keep dipping the forefinger of the right hand somewhat bent into the same dish, which they afterwards thrust into their mouths. The poi which they make from the tarro root tasted to me sour & unpalatable, but when sweetened with sugar it was greatly improved & not altogether disliked by many on board. What most attracted the natives' notice was the unusual large size of our ship to them, which whenever they would come on deck they began sometimes to the number of half a dozen to fathom her length from the stern to the bow along the bulwark & at other times laying themselves flat on the quarter deck. Their knowledge of measurement appeared to us to be yet but very imperfect & only amounting to a calculation of fathoms carefully taken by the hands dependant on the memory, different to the Africans who tally their number by notches on a stick. At 8 o'clock in the evening we was again surprised to hear the eight o'clock gun fire from the fort as is customary with the garrisons of other nations at that hour. The people on shore kept quiet & peaceable all day excepting for an hour or two when those on board went on shore. Thermometer at 6am 72.5 at 12 76.5 & at 6pm 73.5.

7. A fine morning. In the forenoon went on shore with Lord Byron & most of the principal officers on board, where on our landing we was met by Captain Charlton & a number of the chiefs ready for to join us in procession to Mr. Pitt's (Karaimoku's) house for the purpose of being introduced. The chiefs were all decently dressed in the European costume & on our leaving the landing place for Pitt's house they in a friendly manner placed one of us between every of themselves each having hold of our arm till we reached our destination, which was no other than a small thatched hut pleasantly situated near the farther end of the town within a large garden fenced round with high poles standing upright. When we entered his hut we found him seated in an arm chair, dressed in a black robe or gown with breeches of the same colour & had on his feet shoes & stockings. His visage thin and defective of an eye, but by no means unpleasant in his manners. At the upper end of the hut I saw a platform was raised about a foot from the floor covered with the same kind of mats as the rest of the hut, where the young King & his sister brought with us from the island of Mowee were seated on a shabby sofa placed crossways, having at the back of it several handsome large feather plumes of various colours, customary used by them on former times when at war or high festivals. The King was draped in a short, blue jacket with shirt & pantaloons, without any shoes or stockings & his sister in the mourning dress brought her by Madam Boki from England. Both looked to be of a delicate constitution & of rather dark complexion, full nostrils & large mouth, but had fine open countenances, with good eyes & teeth & not altogether wanting a sensibility of look that rendered them engaging. From the King's sofa on the right stood up in a line along side the hut, his principal chiefs & his general officers. From his sister on the left lay on the mats by her sofa Queen Kaumanna & two more of their father's (Tamahamaah) wives that are yet living. After these three stood according to their seniority five more of their deceased brothers' wives, joined by a number of other female chiefesses to the bottom of the hut, draped in black canton crepes or silks of the same colour with shoes and silk stockings & showey combs in their hair, every way neatly draped contrary to what we all expected to find these people. When the ceremony of being introduced to the Regent (Billy Pitt) was over, through the means of a Spaniard (Mr. Marin) who acted as an interpreter. The present brought from England

was then opened & presented to His Majesty by Lord Byron. The presents brought him consisted of a suit of Windsor full uniform dress and an elegant sword[,] also a gold watch, with the arms of the King of England engraved on the back. Pitt likewise was presented with a handsome gold ring. The King immediately after receiving his new suit of clothes tried them on, when they were found to fit his size the same as if he had been measured by a taylor. He looked in them extremely well & felt himself not a little proud when he had on his cocked hat & sword walking about speaking to some of his subjects. After the ceremony of giving away the presents was over, we were all ordered to take our seats on chairs placed on purpose in the middle of the hut, when country wine was introduced for our refreshment that had been made last year by Mr. Marin who acted as our interpreter. While some sat tasting the country wine, others were getting introduced to the females, who had by this time nearly all sat down upon the mats where they before had been standing, which when spoken to could only answer by a smile for want of their not knowing our language. The Queens were all of enormous size & far from being handsome unless they claimed the honour of being such from their great bulk. Their hair was straight & of a brownish black colour, neatly put up after the manner of the Europeans. Being myself tired of waiting any longer at this mock sort of formality which already had taken up several hours, that afforded to others of our party some amusement, I left unperceived to look after plants, but was soon after missed by His Lordship who sent after me to return, when on reentering the hut I was now introduced to the Regent as being the person who brought their plants from the Horticultural Society of London & that I wanted his permission to collect the wild plants of the country for the Society, to which he replied that I was at full liberty to collect what plants I wanted on any part of the islands. His Lordship having requested me not to go collecting till after the funeral was over, I therefore during the rest of the afternoon accompanied by some of my companions come on shore took a walk through the town, which is situated on a sandy flat without scarcely a tree within a mile to form shade excepting a few coco-nut trees together in places along the sea beach. The huts of the natives are small & mostly all thatched with grass from top to bottom, but there are others who are inhabited by some of the Europeans & chiefs which they cover over with mud half a yard thick to prevent accidents by fire, which do so often occur with the others; these are much better finished inside & the floors are generally spread over with mats made from either rushes or the leaves of the pandanus. The huts of the poorer class are mere hovels, having low door places at

Kiaukiauli. King of the Sandwich Islands. From Maria Callcott's *Voyage of H. M. S.* Blonde *to the Sandwich Islands, in the years 1824-1825. Captain the Right Hon. Lord Byron, commander.* Painted by Robert Dampier. Engraved by Edward Finden.

Naheinaheina. Princess of the Sandwich Islands. From Maria Callcott's *Voyage of H. M. S.* Blonde *to the Sandwich Islands, in the years 1824-1825. Captain the Right Hon. Lord Byron, commander*. Painted by Robert Dampier. Engraved by Edward Finden.

either end or front; where they creep on hands & knees to get out & in to them & without any thing to shut them up at nights but a piece of cloth or a mat. Some of these which we from curiosity went into to examine how they looked inside had a hog or two tied up in one corner & in other places we saw dogs nearly deprived of their hair from the effects of mange. The stench arising from allowing these animals to live with themselves in the same hut was so offensive that we felt surprised how they could exist to outlive one night together. There are only yet four or five houses built after the European manner, two of which are of a middling size the one belonging to the American merchant, the other to Mr. Pitt, not yet finished, part of which is at present converted into a guard house for a few naked soldiers who do duty at times by way of mounting guard in front of the King's hut after a peculiar custom of their own, in the following manner, six or eight of these turn out together & form a line (not a straight one) & walking backwards & forwards one after another till they are dismissed, while they are in this way upon duty, he that is last or foremost, [must] ring a small bell which is carried in the hand by way of signal for the others to turn either backward or forward. Their accoutrements are not all alike. Some have only a bayonet in his hand held upright or reversed just as it suits his own convenience, while another is seen carrying an old rusty long barreled musket of American manufacture without a flint & sometimes a lock, but has at times a bayonet on the top. Some have a cartouch[e] box tied on behind with a piece of untanned goat skin, others have it before & some have none at all. The town of Hana-rura contains from 5 to 600 houses & if the number of its inhabitants are taken at the moderate calculation of ten to a hut where they generally live together in families amounting to two or three generations they will amount in population to nearly the number of 6000, which I think is underrated. Their huts are built without any regular form, enclosed with low mud walls with a small garden, but without tarro ponds the same as we saw at Mowee. Some of these are cultivated with tobacco, Indian corn, water melons & pumpkins &c, while others suffer weeds to grow & neglect cultivation to the preference of a stock yard for hogs, goats, dogs & poultry. Beside the harbour they have built a fort of mud & coral rock picked up in the absence of the tide, it is of a square form & mounted with upwards of 50 guns many of whom are 18 pounders got from the Americans in exchange for Sanders wood. From its situation it is not capable of long defence from an enemy at sea off the mouth of the harbour, it is at present in bad repair & in charge of an old Irishman who had been on the island many years. About a mile above the town on the top of a hill formerly the appearance of a volcano they have another that mount 10 guns which command the town & the numerous tarro ponds with other provisions that are cultivated in a large valley well watered by two rivers which run on each side till joining in one on coming behind the town. Among the many Americans who live on the island, two keep public houses in this place for the accommodation of strangers, where they have already managed to introduce a billiard table each & is supplied with all kinds of spirits as well as wine at times from their country whale ships, touching here for fresh provisions & water. In the evening returned on board with those that had been with me on shore. Thermometer at 6am 72 at 12 76.5 & at 6pm 73.5.

8. (Sunday) A fine day. After dinner went on shore in company with several of the gentlemen on board. Met with Mr. Charlton who appointed tomorrow for me to go with him to Mr. Pitt to arrange matters respecting my going to the woods to collect plants which was now what I anxiously wished for. Among other things which he offered to do for me, was to make use of three of his jackasses brought from England to carry myself & specimens, these [I] assured him would be of more hinderence than of use to me, & that I would rather have the same number of natives in preference. But of this he would hear no denial, till at last I consented to accept of one, with which he seemed satisfied, & parted promising to have what natives I wanted as well as a guide who knew the country from Mr. Pitt tomorrow. With such flattering hopes of success I spent the afternoon idling about different parts of the town along with my companions who did not forget tormenting [me] for refusing the asses which they said would be just the thing for them when they had time to be long enough on shore to go with me on my journey. In the cultivated grounds above the town I noticed during the afternoon some good sugar canes mostly of the purple & striped kind, & a few patches of our potatoes with weak stems, not likely to be productive. At sun-set we all returned on board highly amused with our afternoon's walk which afforded us an opportunity of observing more of the natives' simple manners & their mode of dancing, which they always accompany with a song by graceful motions of the arms & body, raising their voice at intervals to a high key, then again lowering it without any given certain time, that had in the last resemblance to music. They all at this place approached us every where we went with freedom & not in the least intimidated like those we saw at the other islands, likely arising from a more intercourse with foreigners than them. Several we met could speak a good many words of English, to whom they are very much attached & often

heard to make use of the word Briganie Maithi which means English very good & superior to people of other nations, whom they esteem with jealousy. Thermometer at 6am 72.5 at 12 76.5 & at 6pm 73.5.

9. A fine morning. Went on shore in the forenoon to go with Mr. Charlton to Mr. Pitt to arrange matters with him respecting my going to the woods as was proposed yesterday, but owing to the preparations making for the funeral, got nothing done. Had dinner at one of the shabby inns on shore kept by an American of the name of Novara, an obliging disposed man, who is besides an innkeeper an auctioneer & commission merchant, which brings him the customs of all masters of vessels coming to this part, especially that of his own countrymen who enjoy the whale trade of these islands, loading several vessels annually with Sanders wood for which they give in exchange European & Chinese manufactured goods with every advantage to themselves. So eager I was informed have they been since the discovery of this native wood by one of themselves in 1810 whose name was Brown, that every scheme has been entered into, supplying the government with such articles as they thought would attract their notice most to be likely purchased. To gain this end they have introduced slight vessels built of fir, fitted up with showey cabins, such as looking glasses [&] sofas with red Morocco cushions &c. Sold at the enormous price of 80000 dollars to be paid in Sanders wood, as it could be got at the different islands without any farther trouble to the natives than bringing it to the sea beach as they employ themselves small crafts, constantly running among them from the harbour of Hana-rura, where they have on shore large store houses for the convenience of loading vessels. Besides vessels they have furnished them with all kinds of naval & military stores in abundance, also small wooden frames of houses made in America, as well as light carts, some with two & others with four wheels. Nay farther these scheming speculators have lately brought over a [hansom] carriage in expectation of bartering it to advantage, but in this they have not yet succeeded, although the Regent & some of the Queens were indulged to have a ride in it some distance out of town shortly after it was landed. They still prefer the carts where they can lie down with more ease, when they go to bath, which are used rarely on any other occasion, & are always drawn by natives in preference to horses or mules of which they have several that are often rode by them either [with] or without saddles. Bathing alone may be said to induce the higher ranks of these people who are yet but partly civilized savages to leave their homes, where they spend most of their time either sitting or lying down asleep upon

mats, excepting when they have a desire for this favourite amusement which the whole tribe are so exceedingly fond of that seldom during the day the sea shore & the harbour is seen without numbers of both sexes mixed together swimming with perfect ease as if they had been of the same species with the aquatic feather tribe. The Americans although they have been on these islands for years enjoying lucrative trade, that has obtained to many of them sufficient to live with comfort in their own country, it appears none of them has even yet possessed lands of their own in this country whither from aversion of jealousy to strangers on part of the chiefs which is [predominant] among them, or from the more lucrative gain on the part of the Americans to procure Sanders wood only that grants or leases of land have never taken place, which might by growing cotton, coffee or sugar be soon with little expense turned to equal advantage with the other. Be this as it may, they have however but little (if any) improved the morals of the natives & entirely neglected to teach them the arts of agriculture beyond what they understood of themselves when visited by Cook & Vancouver that of growing the tarro root which is their chief subsistence, & even this I was told had been much neglected during the late King's reign who contrary to his father (Tamahamaah) encouraged free permission to gambling & all sorts of vice to the natives. It is said of this King that during the time he had reigned till his departure for England that he was given to every kind of debauchery & so much addicted to intemperance from excessive drinking before he left the country, as to be for months at a time rarely even sober, & while in this state often seen one day naked with only the maro on, & on the next perhaps the best European dress. At sunset while waiting for the ship's boat usually sent on shore at this time for the convenience of persons on shore belonging to the ship to return on board, I met with the Irishman who had the management of the guns of both forts at this place standing in the door of a thatched hut within a few yards of the harbour, which during our conversation he informed me was the King's residence when he left for England & on my telling him that I was surprised he lived in such a poor hut, he requested me to go inside & then I would find it was the best in town. After I had got within I noticed that more than one half had been divided into separate apartments by low canvas screens, which on enquiring what he had these little rooms I was told they were for his six wives who lived separate from each other & never associated together or with himself excepting, one at a time, in another apartment which I saw near the door, now converted for the depositing of naval stores. This elderly man further informed me that the King had similar accommodations fitted up on board a fine brig (*Albatross*) where he at times went with his wives to sea for two months together, taking himself with him as his captain, during which time he had hardly ever been free from intoxication. When I returned on board I found the ship's company had been clearing the hold to get at the coffins, which will be got on deck tomorrow. Thermometer at 6am 72 at 12 76.5 & at 6pm 74.

10. A fine day. In the morning both the coffins were brought on deck, & the lids of the outer cases opened for those on board that had a wish to look at them. Both were covered with fair red cloth & richly ornamented with gold gilt plate, having the names of the deceased engraved on the top. At noon some light showers of rain & at 3pm arrived Mr. Charlton's cutter which we gave up towing sometime after our leaving the Galapagos islands to find her way here herself. During the afternoon went on shore to see Lord Byron & Mr. Charlton to inform them the necessity of my getting a hut some where on shore as I could not do any thing to my plants on board for want of having accommodations, which they were both of the same opinion & promised to procure me one. At sun-set returned on board without getting any farther done, but learned the funeral would take place early tomorrow forenoon. Thermometer at 6am 73 at 12 76.5 & at 6 74.

11. A hazy morning with light wind. At 9am the coffins of the King & Queen were lowered over the ship's side into the launch & shortly after orders was given to most on board to have themselves in readiness by 12 o'clock. At one pm all those going on shore got into the different boats assigned for them, who then formed a line ahead of the launch to tow her with the bodies on shore accompanied by Lord Byron in his gig on the right, & by the first Lieutenant in the whale boat on the left, with colours in all the boats hoisted half mast. The boats on leaving the ship 26 minutes guns was fired, which was continued by the same number of guns from the forts on shore & a few from an American ship in the harbour. The procession formed by the boats while towing the bodies on shore had a grand & solemn appearance. On landing the bodies they were placed on two light four wheeled wagons, where a frame of wood had been placed above the sides to support the coffins, arched over, covered with black native tapa cloth, left exposed behind to have the head ends of the coffins in view & drawn beside each other slowly on by their late Majesty's favourite domestics, having before them our marines & the band playing slow marches, & after the wagons joined next to the bodies his

present Majesty supported on the right by Lord Byron & on the left by Mr. Charlton the British consul, both in their full uniforms & the King in his new Windsor uniform. After these three followed our Lieutenant & their other gunroom messmates, joined by a number of Queens & High Chiefs with their wives & all dressed in black after the European manner, but many of them without shoes or stockings. Next to these again came myself with some more steerage passengers, intermixed with several natives of a more inferior rank, followed by about 100 sailors cleanly dressed forming a procession of four deep with midshipman in full uniform on their right & left. In this manner we kept slowly advancing between a line of native soldiers standing at short distances from each other, with their rusty firearms reversed & most of themselves naked with only the maro on. Some few of them we saw had on Russian military jackets, that gave them very ludicrous appearance, from the rest of their body being exposed naked, but that which the jacket covered. One singular from the rest attracted most of our notice from his having on an old black hat & a white dirty shirt half way down his thigh, leaving the rest of his legs bare. Both of these articles are seldom seen worn by the inferior [class] of natives. After we had passed the man in the shirt then we came to 12 (six on each side) in handsome coloured feather tippets, which we concluded to be of a higher rank. Before the band was carried five large beautiful feather plumes mentioned to have [been] seen behind the King on the day of being with others introduced to Pitt. These were carried in a leaning posture as is customary with them when going in battle, which from their variety of colours added much to the solemnity of the scene on the occasion. On our way we halted outside the missionaries' chapel, where in front they have a bell hung on an old frame of wood. Here our chaplain read over a prayer in English and after him one of the missionaries another in the native language, which when done we again resumed in procession as before, till arriving at Mr. Pitt's hut between two & three o'clock, when the marines formed outside a line in front. The weighty coffins were now taken from both the wagons by the sailors & deposited inside Pitt's hut at one end where a low platform of wood had been previously erected on purpose covered over with fine mats of various striped colours & hung round above with black Jap cloth. After the bodies had been safely deposited under the same roof with Pitt himself. The Queens of the deceased joined with three of his father's still alive & who are of a higher rank stood at the head of them in forming a line on our side the hut joined by the wives & daughters of the chiefs, the same as on the 7th & on the opposite side, Pitt, Boki & other chiefs of consequence with the principal officers of the *Blonde* & a few Americans, no inferior class of natives being admitted, or before allowed to join in procession. We had not long entered the hut when the chaplain gave out an anthem which was sung accompanied by the band, then he afterwards read a prayer, & the missionaries in a similar manner went through the same form in the native language. Pitt who all along never left his hut to join the procession, remained at home quite composed in his arm chair draped in his usual black gown. The whole of the funeral procession throughout commencing with the disembarkation of the bodies from the ship till deposited, was conducted with solemnity in a style far superior to what most on board before going on shore had even expected could be done. But notwithstanding there was one fault allowed to occur to any observer like myself free from partiality, that in not giving a general invitation to the respectable few Americans resident in the place, or even afterwards when they attended at the landing of the bodies from the ship, no invitation was given them to join us in procession, but suffered to form unasked a small body of themselves at some distance apart on one side all the way. What might be the reasons assigned for treating them with such distant marks of disrespect is best known to those who had the conducting of the funeral. It could not be from any wish on the part of the chiefs, for they were often seen on the way to leave their places, particularly Boki, & join the Americans with whom they appeared to be on the most friendly terms. When the funeral ceremonies was over all persons belonging to the ship were ordered to return to their respective boats in marching order with the band in front playing lively airs, which brought nearly all the natives in the place from their homes eagerly running towards us like mad people on hearing the sound of the great drum that so much astonished them, they regardless of their own soldiers broke instantly on both sides through their lines & came in amongst us all the way till we reached the harbour to embark, which afforded to many, particularly the sailors, good amusement. Before embarking the ship's company had a glass of grog each & the officers a similar refreshment at His Lordship's house. At 5 o'clock we all arrived without any accident on board, farther than a few of the marines managed to get a drop too much. Mr. Dampier the draftsman had been at an early hour stationed in some convenient place, where he could with best advantage take a drawing of the funeral procession from the disembarkation to the place of internment. This gentleman has already I am told commenced taking portraits of some of the Queens, with whom he has taken up his residence among them at their homes, some say for the purpose of getting a more correct painting. Thermometer at 6am 73 at 12 76 & at 6pm 73.

Native Girl of the Sandwich Islands. From Maria Callcott's *Voyage of H. M. S.* Blonde *to the Sandwich Islands, in the years 1824-1825. Captain the Right Hon. Lord Byron, commander*. Painted by Robert Dampier. Engraved by Edward Finden.

12. A fine day. In the forenoon had the permission of Lord Byron to go on shore to collect & also the favour of a lad from the ship to assist. At 3pm took some salt provisions & some biscuits, with my bedding &c. on shore, which for want of a hut of my own got liberty from His Lordship to put them for safety at present in one of the rooms of a wooden house which had been given him during his stay. Went afterwards to the hut of one of the natives (Manaware) brought with us out from England & who had often offered me when he had got home to consider his house my home, but on my now reminding him of what he had promised, he treated me rather cooly & forgot the obligations, which he repeatedly owned to be indebted to me on the passage. He said that he could not accommodate me in his house, nor did he know where a hut could be got, but if I applied to Pitt tomorrow he would likely find me one. Being this unexpectedly disappointed after so many promises given before, I left him to go on board Mr. Charlton's vessel in the harbour, where he lives in preference to his being on shore, but here I met with the same disappointment, from his having Lord Byron & several others with him at the time to dinner. Having determined to go early in the morning to the woods, I sent the lad on board Mr. Charlton's cutter to boil a piece of salt beef for our journey, having already lost too many days idling on shore. In the evening met with Mr. A. Bloxam who on finding that I was going to the woods tomorrow, wished to accompany me & proposed we should both take up our lodgings in his Lordship's house for the night, & then early in the morning we could go away together. Thermometer at 6am 70.5 at 12 74.5 & at 6pm 72.

13. A fine morning. Got up at 4am & called the lad (Mantle) who I left on board the cutter all night to prevent his being out of the way this morning. By 5 o'clock we was ready to proceed on our journey, taking with us some salt beef & biscuits. Staying shortly after on our way where Charlton kept his asses, for to take one of them with us, which after considerable delay putting on the pack saddle, on the top of which we secured the heaviest articles of our traps, then gave the ass in charge to Mantle, who could neither ride, lead, nor drive it before him fast enough to keep up with us, walking slowly, which put the poor fellow out of temper sadly & at last on getting to the tarro ponds about half a mile above the town we found that considerable delay would occur going some distance round to get the donkey farther on our journey. It was therefore agreed upon amongst us to turn him a drift & to leave the pack saddle (mostly goat skins) & bridle in one of the native's huts. These tarro

grounds had loose stone walls round them outside for some miles in length. While we crossed through them to get towards the wood at the head of the valley, I noticed no difference in their cultivation from those seen at the island of Mowee, but that they were here rather better kept weeded. The ridges between the ponds had growing upon them abundance of sugar canes, besides other articles of introduction, such as the tobacco, cabbages, water melons, pumpkins &c. After having travelled about three miles we came to a fine clear stream of water at 8 o'clock, where we halted to have some breakfast, & had scarcely made a beginning, when crowds of the natives came from the neighbouring huts where we was, & soon after formed a ring sitting on their hams with us in the middle, all the while staring on us with surprise while eating our mouthful of beef & biscuits in a manner compared to so many starved dogs, which in hopes to shame them away a little farther from us we freely parted with some of what we were eating among them, but instead of having the effect we wanted it only made them encroach upon us with more freedom, till at last we was under the necessity of tying what little provisions we had as their numbers were so fast increasing that we in a manner felt doubtful from their ravenous appearance, but they might attempt to deprive us of the whole we had. During the time we stopt by the river I gathered two species of *Polypodiums* & three of *Convolvuleses*, besides a *Hibiscus* with flowers white inside & light purple out. On resuming our journey among the tarro ponds followed by a number of natives, we had not got far when we met with a chief getting up sweet potatoes, who addressed us in English very friendly & who on being made acquainted with the objects of our journey willingly gave a young active boy to Mr. Bloxam to carry his bag & other traps till we returned to town on being informed he wanted one for that purpose. By 10 we began to enter the woods, having travelled about five miles most of the way through tarro ponds & cultivated patches of sweet potatoes, bananas, water melons &c. & in places near the river grew on each side a few low unhealthy bread fruit trees not in bearing. The tea tree (*Dracaena terminalis*) of the country grew in all places uncultivated in abundance, from two to four feet in height & about an inch & a half in diameter. Beyond the provision grounds where it grew with such abundance we observed several large sheds where the chiefs had formerly during the last King's reign, been in the habit of distilling great quantities of spirit from the fermented liquor made from the roots, by means of large iron pots, obtained from whale ships in exchange for provisions. We had hardly entered the wood when we met with two trees of the *Eugenia malaccensis*, with a number of birds sucking its red blossoms, one of which Mr. Bloxom had the luck to shoot, which he said was a species of humming bird. My meeting with this species of rose apple to every appearance growing here indigenous so far from habitations rather surprised me to have met with in the Sandwich Islands. The other trees seen most common as we continued travelling on, were [*Aleurites*], & a species of *Acacia* used by the natives for making their canoes & paddles & on the ground underneath their sheds as well as upon the trunks grew mixed together a numerous variety of handsome ferns in all kinds of situations either moist or dry, covering the surface almost every where met with. On our coming more towards the center of the island the trees became more lofty & the ground underneath more shady & damp, when there appeared several species of *Psychotrias* and *Beslerias* with two or three tall growing *Lobelias* producing splendid clusters of flowers, as also three kinds of *Metrosideros* with rich bunches of scarlet flowers, having about them a number of red birds sucking the honey from their blossoms which we shot but could not find them afterwards, owing to the number of ferns & other plants on the ground underneath. At five in the afternoon we gained the head of the valley, having travelled about 10 miles, when we now had the pleasant view of seeing the Pacific Ocean on both sides the island, which afforded us a very gratifying sight. Here we halted for half an hour to enjoy the cool breeze which was occasioned by the valley terminating in a narrow gap, overhung on each side by high abrupt cliffs covered with tufts of *Lycopodium cernum* so common within the tropics, and two or three kinds of *Vacciniums*, besides other low shrubby plants which gave this mountain to its summit (at least 3000 feet high, 1000 of which the path leads over to the valley below on the north side of the island) the pleasant appearance of verdure. During the time we remained at this place which is a thoroughfare for the natives crossing the island from one side to the other, we was agreeably surprised to be overtaken by [a] man somewhat better dressed than the rest of the natives who accosted us in English & very familiarly of himself entered into conversation giving us part of his history, such as being a native of Otahite where he had left when a boy as a sailor on board a whale ship, then afterwards served in the British navy till he got wounded at the Battle of Algiers, when he was discharged unfit for service with a pension of £25 a year. Having ascertained of this Otahitian that we could travel by the sea coast round the east part of the island to town during the next day, we now made up our minds to remain at some hut on this side for the night, that we might succeed the better to undertake so desirable a journey back to town, but to advance farther on our journey we had yet the most difficult task to perform, before we could descend to the bottom of the precipice below by a winding rocky path in places for several yards together quite perpendicular, that required us to take off our shoes to prevent any false step & to scramble the best way we could at times backwards on our hands & knees, when the natives in our company carried their loads seemingly with perfect ease & enjoyed heartily, when they saw us in the least terror to go on & have to come to our assistance to prevent our falling. After we succeeded in getting to the bottom fortunately without any accidents, we soon after came to two or three small huts, where we with little trouble obtained lodging for the night in the hut of a Bengal black, who informed us, [had] been wrecked in an Indiaman on the coast some years ago, & has since lived on the island with a native woman by whom he has several children. This fellow who was very communicative & spoke good English, informed us that he followed the tayloring trade & sometimes had acted as cook for the King before he went to England. Having yet some hours of day-light before us, we proposed to have a mouthful of beef & biscuit, & then afterwards to walk to the sea side below about a mile & a half distant, while the Bengal black promised to have something cooked for our supper when we returned. On our way to the sea side we passed several of the native huts with little patches of cultivated ground attached to them, chiefly planted with *Broussonettia* which they from the bark make the tapa cloths. They also cultivate another plant which they name <u>Nane</u> [*noni*] in the same manner as they do the *Broussonettia* for the sake of it which yield their favourate yellow dye for their tapa cloths. Besides these we noticed in the ravines we had to cross some tarro ponds & several groves of healthy trees of the rose apple, growing seemingly without any cultivation & coming into fruit. The outskirts of the bay is full of fish & tarro ponds around which are a number of natives' huts thickly inhabited. The bay is open & exposed full of rocks in many places above water which renders it unsafe for vessels coming to anchor. The high ridge of mountains, which run nearly central in the island terminate on this side (north) in abrupt cliffs & along the coast for four or five miles back in the interior, is as in the neighbourhood of the town of Hana-rura, left everywhere uncultivated the same as nature had formed it, excepting the small patches that surround the natives' huts. Although the soil is good & producing excellent pasture, none of the government cattle have yet been sent on this side from Hana-rura to feed. When we returned to our hut at dusk we found some tarro root ready cooked, or rather baked under ground with hot stones, & a small fresh water fish of the mullet kind cooked in a similar way enclosed in the leaf of the tea tree. This with a little of our beef & biscuit made a comfortable supper & about an hour after we went to sleep on mats by the side of the Bengal black & his wife, with her father & mother & the rest of the family in all upwards of a dozen, men, women & children, besides several dogs, thickly stowed together at one end of the hut without distinction. Thermometer at 5am 74 at 12 78 & at 7pm 74.

14. A fine morning. Got up at 4am after a restless night from fleas & cold, the hut being so open that light could be easily seen through it, in most places. Having ascertained from our landlord last night & this morning that if we intended going home round the island by the coast, it would take us two days instead of one as we had been informed by the Otahitian, which for want of provisions enough for the journey we was reluctantly obliged to return again to town by the same road we came & at five o'clock while the heavy dew yet remained on the grass we parted from him, after giving him a trifle for his last night's trouble, when we then proceeded on our journey towards the steep rocky precipice found yesterday so difficult in passing, which on our reaching the path at the bottom, overtook numbers of the natives who commence early in the mornings to travel on account of the heat after sunrise. Most of them were all loaded with provisions for the chiefs, such as hogs, tied by the four feet together carried on the back & at times head, or drove before, while others carry bundles of tarro, or large calabashes of poi secured to each end of a staff thrown across the shoulder, the one hanging before & the other behind, keep travelling on in this manner with a short shuffling step with perfect ease, stopping to rest themselves at every other short distance to have a fine draw in turn from a wooden tobacco pipe, then they resume their journey as before, having no calculation of time beyond that of the rising & setting of the sun to hurry them on farther, than to gain some hut for to put up at for the night, being afraid owing to superstition to travel after dark. Having had more opportunity during to-day & yesterday than hitherto of seeing the manners & customs of these people form curious little parties carrying their provisions to the chiefs or to market. Both sexes are generally tatooed irregularly, some with figures of goats, muskets, & even letters of the alphabet, name & birth place with the date of the year is not unfrequently seen along the arm. The men many of them shave their heads & cut their hair in the form of a helmet, the crest of which is often bleached with lime, so as to be of a light whitish colour. The women esteem it cut short, keeping only a rim over the fore-head bleached white & standing in front like bristles,

sometimes a long curl is preserved on the middle of the fore-head, which is combed backwards. Some suffer their hair to grow & tie it up behind in a bunch or queue. Many of the females different to any of the other sex, have a tatooed line about two inches broad inside the thigh down as low as the ankle, where it terminates in that part in form of a ring generally on the right foot. Most of them had a looking glass & a wooden tobacco pipe tied round the neck in a handkerchief or a piece of tapa cloth. So great is their desire for smoking, that they are seldom seen any length of time without using it & what is something curious to observe among them that although most of both sexes carry their own pipe and tobacco, rarely more than one at [a time] is seen to light their pipe, which after a few draws of smoke taken first by the person who filled it. The pipe is then handed to the next, & so on till the whole present are satisfied without any distinction of persons. By 10 o'clock we succeeded in getting homewards nearly through the wood, when Mr. Bloxam got impatient to wait any longer with me & soon after parted to be home before the sun's heat increased, while I remained for some hours longer behind collecting more plants not met with yesterday. When [I] arrived in town at His Lordship's house I found Mr. Charlton had not provided any place for me to live on shore since I left. I was therefore reluctantly obliged to go on board to lay my specimens of plants in paper, but for want of boats being regularly sent on shore, I did not get on board till nearly sun-set, when Mr. Charlton kindly favoured me with one from his own vessel, who at the same time promised to have [a] hut for me to live on shore tomorrow. Thermometer at 5am 62 at 12 78 & at 7pm 73.

15. (Sunday) A fine day. At 10 church service performed as usual. Employed most of the day laying in my specimens, sorting seeds, & making more seed papers. Before church time the captain enquired how I thought the hut that Mr. Charlton had got for me would do, as he wished every one of us to be comfortable, to which I replied it was good enough, but that I was informed it was empty & would require a couple of stools & something by way of a table to write upon, which he immediately ordered me to have or any thing else I thought he could do for me. At 5pm the whale boat going on shore with the surveyor, purser & a midshipman got upset in the surf on the reef outside the harbour & remained in the water upwards of 20 minutes till another boat who had left the ship shortly after came to their assistance & picked them up nearly drowned holding on by the keel of the boat turned upwards. Thermometer at 6am 73 at 12 76 & 7pm 73.

16. A fine day. At 10 o'clock went on shore with my friend Mr. Forder who intends to remain with me to amuse himself in drawing plants for me & to take of the hut while I am absent in the wood. By 2pm got all my traps removed from the landing place to my hut, which is about half a mile distant east of the town & has never been tenanted before. In the afternoon shifted my specimens, while the lad Mantle hung our hammocks across the corners of our hut & set what little furniture we had to rights as he thought proper. Before dusk Lord Byron & the surgeon rode out to inform me that they were going early tomorrow with a party to the Pearl River by water & that if I felt inclined to accompany them I would better come to breakfast with them at 8 o'clock, which on returning His Lordship thanks promised to be with him at the hour appointed. Thermometer at 6am 72 at 12 76.5 & at 7pm 74.5.

17. A fine day. Got up at 5 o'clock & wrote till 8 when the lad Mantle & myself left for Lord Byron's house taking with us the necessary traps for collecting as well as some salt beef & biscuits. But did not embark before 12 at noon. The party consisted of His Lordship, surgeon, draftsman, both the Bloxams, Mr. & Mrs. Charlton & sister including myself a master's mate & a native lad commonly called Sir Joseph Banks as our pilot. After two hours sailing along the coast we entered the mouth of the Pearl River, which divides itself into several branches forming two islands, one which is smaller than the other is called Rabbit island from a person the name of Marin a Spaniard residing at Hana-rura having put rabbits upon it some years ago which have since increased to numbers. We had scarcely entered the river when it became nearly calm, in consequence of which His Lordship with Mr. Charlton & the Bloxams left us in the launch to row a small whale boat we had towing after us, & soon pulled ahead out of sight, which occasioned much delay afterwards, partly from our being strangers to where the several branches of the river [led] to & the stupidity of Joe Banks, not telling us where he was ordered by Pitt & Boki to take us all to for to remain for the night. Hour after hour passed away in useless suspense, the one party pulling under a hot sun in one branch of the river, while the other kept sailing & tacking about from bank to bank in one or two different others in expectation of being seen or overtaken, by those absent, till at last the one hauled their boat on shore & the other dropped their anchor opposite to each other on Rabbit island, which both were at the time ignorant of till observing each other shortly after strolling about on the island. The only hut upon the island happening to be near at hand those of the party on shore walked towards it, while the launch

having the ladies & some others on board got up the anchor to sail round to the hut, where they all landed with help of canoes at 6 o'clock somewhat discontented, particularly the ladies, but after having sat down to dinner on the mats spread out on the grass, matters were soon again restored to perfect harmony & the wine bottle passed round freely to every one of the company to drink each other's health without form or any kind of ceremony in respect to persons. At dusk we again embarked to cross the river to the opposite side to remain at a larger hut for the night, where we arrived at 8 & was soon after landed by the help of the natives' canoes. At 10 o'clock two old men entered our hut to play the hura dance on a couple of large bottle shaped gourds in a sitting posture beating time upon them with the palms of their hands accompanied by a song made upon their late King. The hura dance is always accompanied by graceful motions of the arms & body, with the same spirit, standing or sitting. About 11 we all retired to rest lying down beside each other on mats, some with pumpkins or what else they could get for a pillow. The ladies got themselves screened off in a corner with a flag beside us without any other accommodations. Pearl River is situated about 7 miles west of Hana-rura & is very improperly called a river, being rather inlets from the sea branching off in different directions. There are three principal ones named by the surveyors: the East, Middle, & West Locks. The entrance to it is very narrow & shallow & in its present state is only fit for very small vessels to enter, but over the bar there is deep water & in the channel leading to the Locks is from 20 to 7 fathoms, but the Locks themselves are rather shallow. The sea coast from Hana-rura for some distance to the westward of Pearl River possesses no variety of plants, beyond two or three species such as [*Argemones*], *Portulaecas* & a few other little annuals intermixed with the common long grass, so plentiful on the coast every where round the island. The oysters that are found in Pearl River are small & insipid to the taste & of no value or consequence. Thermometer at 6am 74 at 12 84 & at 7pm 82.

18. A fine morning. Got up at 4am after a restless night tormented with fleas, & shortly left with my man Mantle for the woods leaving the rest of the party yet asleep, but on travelling about three miles the path which we at first got into terminated, when afterwards the grass became every step much longer & difficult to travel, till at last on getting other three miles farther we got so much entangled with creeping plants running a little above the ground underneath the grass, that Mantle for want of having stockings on his feet, sat down shedding tears complaining of his ankles & refused to go any fur-

ther. Being yet more than five miles from the woods & our provisions not sufficient to last for two days, we was therefore obliged to return to town some distance inland from the coast by a path leading through tarro ponds & uncultivated grounds. On our returning to the path which we had left near the Pearl River, we came to several huts thickly inhabited on the sides of a ravine plentifully stocked with bananas, tarro & other vegetables, as well as several healthy bread fruit trees, just forming their fruit. Here we met with an old Englishman who informed us that there was on the opposite side of the ravine a large river coming out under ground, which on getting to the place, we found to our surprise what he had told us was correct & remained for some time admiring so curious a subterranean stream of fine cool water. Its source was rapid forming a cascade nearly 20 feet in height, having on its sides some ferns & mosses. In the grounds of the natives I was shown plenty of the ava plant (*Piper*) mentioned in the history of these islands, as being destructive of the health of the natives when used to excess from its intoxicating quality. Of this plant I obtained several specimens in flower. During the time I remained here I learned from the poor man met with, who is now old & infirm, that he had been on the islands upwards of 16 years & that the grounds which we were now upon belonged to Boki, which till lately he had under charge for 10 years past, but on Boki going with the King to England another chief turned him away & taken all his little grounds from him, which has since obliged him to live on the charity of the natives. The neighbourhood of Pearl River is very extensive rising backwards with a gentle declivity towards the woods, but is without cultivation, excepting round the outskirts to about half a mile from the water which is divided into separate allotments or farms belonging to the chiefs, generally enclosed with walls from four to five feet high made from a mixture of mud & stone. On these farms live the poorer sort of natives & often a few wretched ragged foreigners, who have a hut given them with a small spot of ground for which they must work a certain number of days in the week for the chiefs, besides paying an annual rent in dogs, hogs, goats, poultry & tapa cloth, which they are obliged to carry to whatever part of the island their master may be living & on the least neglect to perform what is demanded from them, they are then turned away & deprived of what provisions or stock they may possess, by the covetous & indolent chiefs, without any respect whither natives or foreigners. Such is the present despotic, or absolute laws in the Sandwich Islands, corroborated by all the foreigners met with at different times, who on our arrival hoped His Lordship would render them, their little property more secure in future as was reasonably expected, but to their misfortune they must yet remain in their present condition, till something is done for them by the British Consul, as we have no authority as far as I could learn to interfere with the laws of the country. As we travelled homewards the country on one side towards the woods still remained as nature had left it, & the same on the other towards the sea coast, excepting in places where we had to cross now & then the bottom ends of ravines cultivated chiefly with tarro in ponds, which resemble much peat mosses that had been worked & get afterwards full of stagnant water, which while they continue such a system of cultivation without being confined to certain limits, no convenient road will be made to travel, even on foot any distance on the island. At the side of one of these ravines which we crossed, we met with another subterranean river much larger than the other, but of no great fall after its appearance from under ground. By 4pm we gained the summit of a high hill thickly covered with tufts of long grass, within three miles of Hana-rura, when a heavy shower of rain fell. On the top of this hill, there is a burial ground of the natives, where formerly the chiefs of rank had for themselves a morai. At the bottom towards the sea their is a circular salt pond nearly two miles in circumference pleasantly surrounded by other low conical hills & in places on the sides of the ravine leading to the pond from the interior, are several huts of the natives with some tarro ponds & a large grove of coco-nut trees apparently very old from their great height & mossy appearance. At 6 o'clock we reached town somewhat fatigued & wet, having travelled since morning about 20 miles without much success, being too near the sea coast, to meet with a variety of plants. However the journey afforded me more information respecting the present mode of life, which the natives live in, & the manner in which they yet continue to cultivate these grounds, differing but little (if any) from the time described by Captain Cook & others. When on our way through the town we learned that His Lordship with the rest of the party had only returned about an hour before us. Thermometer at 5am 65 at 12 83 & at 7pm 78.

19. A fine morning. In the forenoon put my specimens of plants in paper, that I had collected during the last two days, & in the afternoon shifted those collected on the 13th & 14th. At 4pm light showers of rain. At dusk Lord Byron & the surgeon called to enquire the particulars of my journey home by land from the Pearl River & shortly after came two of the American Missionaries who on going away invited me to call upon them when I could make it convenient. The lad Mantle still complains of his feet, but promises to go

with me in the morning to the woods. My good friend Mr. Forder is busily employed figuring the kinds of *Portulaecas* brought home yesterday from the hill above the salt pond. Thermometer at 6am 75 at 12 86 & at 7pm 77.

20. A showery morning. Got up at 4am & at 5 had some tea ready, when shortly after Mr. Bloxam came to accompany me to the woods. The morning getting fine at 6 o'clock we left the hut, but had scarcely entered the wood, when at 8 Mr. Bloxam said he shot birds enough to skin during the rest of the day & would now go home to get himself shifted of his wet clothes, being rather surprised at his intentions of going home so soon, I asked the favour for the first time to shoot me a few birds as he had enough himself, but this he refused giving me as an excuse that he intended all his duplicates for Lord Byron, & then bid me good morning. Shortly after he had left, Mantle taxed me with too much liberality towards him by giving him the use of a small tin box during the voyage to carry home any small birds he might kill, observing at the same time, that the excuse he made was no other than a denial to oblige me. We had no sooner entered the wood, but we met every step as we kept advancing a multiplicity of ferns, many of whom were different to those seen on the 13th & 14th in Hana-rura Valley. By 12 o'clock we gained the summit of the highest hill fronting the bay, while it rained in torrents & not a dry stitch of clothing on our bodies. Here I took the temperature of the air which stood 69. This hill is upwards of 2000 feet above the level of the sea, but there are others beyond it, central in the island of much greater height. On travelling two miles beyond this high hill the wood became difficult to penetrate, besides very disagreeable owing to the rain, & having by this time loaded ourselves with plants, we now took a mouthful of spirits & a little biscuit, which helped to remove the shivering fit upon us & then we returned slowly towards home. The number of different ferns met with during the day exceeded 12 & the variety of other plants considerable, among whom was a parasitical species of *Metrosideros* with large scarlet flowers & two or three of *Lobelias* with large clusters of handsome flowers. I met also for the first time with small land shells, having a variety of rich striped colours. These shells were found chiefly on the leaves of the wild tapa (*Broussonetia* spec.) of which there are several species with different coloured flowers. On our coming to some huts a short distance from the wood we saw several of the natives eating sweet potatoes raw, which had been newly taken up & not even washed free from sand. At 6 o'clock got home to our hut & had scarcely shifted ourselves when His Lordship & the surgeon called, as they were riding farther

on towards Diamond Hill, to request that I should meet them tomorrow forenoon at Pitt's to see the plants brought out from England properly transplanted. Thermometer at 5am 70 at 12 at home 86 & at 7pm 77.

21. A fine day. Employed till 12 o'clock putting specimens of plants in paper, then went to meet Lord Byron according to promise at Mr. Pitt's for the purpose of transplanting the fruit trees brought from England, but on getting to Pitt's found he had not yet been there, which after waiting for some time in expectation of his arrival, I had to call upon him at his house, where I met Mr. Charlton who had doubts on his part if any thing could be done till late in the afternoon, owing to Pitt & the other chiefs being now asleep, who would not get up under any necessity whatever. Having informed both that I was anxious to go to Diamond Hill this afternoon, they recommended me to go by all means & not to wait for the certainty of being able to finish planting out the trees. By 3 o'clock Mantle & myself had dinner, then we set out for Diamond Hill & on our way was joined by a native, who offered himself as a guide, & at half past six we with some difficulty gained the summit, which is high & steep without any thing growing on its sides to hold by, excepting short tufts of dry grass bedded in loose sand, which came up with ease & occasioned our ascent the more difficult. Diamond Hill forms a promontory point, or head land, East of Hana-rura about three miles near the sea, & has at some former period been beyond doubt an old volcano crater, now closed. In the center is a level flat at least two acres in size covered with longer grass than the external declivity, on the inside which is next the sea the depth is upwards of 500 feet, below the narrow ridge round the top, which is almost circular. Some parts of the rocks had the appearance of common quartz, others resembled burned lime stone which were soft & white in colour. Along the coast on both sides of Diamond Hill are [a] number of the natives' huts pleasantly situated underneath the shade of banana & coco-nut trees, but in the ravines towards the interior (which are chiefly of volcanic origin) they are thinly scattered at the bottom of the ridges on each side & without coco-nut trees which are seldom seen to grow far from the sea coast. Near Hana-rura is a similar crater, now a fort with several guns placed along its ridge at certain distance from each other, which at the time we first made the island on firing a salute with the other fort near the harbour had a pleasing effect. While returning home along the narrow ridge on top of Diamond Hill, where it is very abrupt & rather difficult to pass. The wind at the time blow rather strong which while picking up some specimens of stones carried away my hat

over the outside declivity towards the sea, when Mantle immediately went after it, at the nearest place he could descend of his own accord contrary to what I wished, as it was now getting dark & the native with us would bring it me in the morning. The native who remained with me while I was getting down to the bottom had scarcely ever uttered a word since Mantle had been gone, till after waiting for some time on getting to the foot of the hill, Mantle in the dark began to call out aloud for us not to leave him behind & running towards us much frightened, which now the native perceived & laughed heartily holding out his hand in the direction where Mantle was coming from, using the words "houry maiti" (no good). When Mantle came he told me the cause of his fright, was the discovery of having picked up a large human skull in the place where he found the hat, but being in a hurry & dark he did not perceive what he had got till thinking he was near me, when he then looked at it & saw the hollow parts of the eyes, which made him call out to us as he did & then unfortunately threw it away. But on enquiring whether he thought he could find it for me again, he said it made no difference as the same place where he found it was full of others, which on account of our having a native with us, we durst not venture on taking any away. At 8 o'clock when we got home I related to Mr. Forder Mantle's fright in going in search of my hat which afforded some mirth at his expense. I learned from an American who had been on the island for six years & now with me since I came on I had hired to cook & make himself otherwise useful that the chiefs used in former times to take their criminals upon the top of Diamond Hill, to put them to death, by throwing them over the precipice where they were left above ground without being buried in the ravine that Mantle met with so many skulls. Thermometer at 6am 73 at 12 86 & at 7pm 76.

22. A fine day. Employed most part of the day at home among my specimens & seeds. In the afternoon Mr. & Mrs. Charlton & sister called upon me to look at my collection made since I came on shore. Mr. Charlton as usual friendly offered me any assistance he could give & promised me when I came to Otahite he would have best accommodations provided for me. Thermometer at 6am 74 at 12 86 & at 7pm 76.

23. A fine day. Employed shifting & numbering specimens of plants & seeds. At 4pm called the Governor of Attooi, lately raised to that honour from his heroic bravery in quelling the insurrection which took place on that island about two months before our arrival, for which besides Governor he is styled

with the honour of being named Duke of Wellington the conqueror of Attooi. He is a young man about 25 years of age, stout made & rather good looking in comparison to the rest of his countrymen. In his youth he was brought up by Boki, who was in England & until lately made a General & Governor, went under the name of Tamaahamahaa. He had not been long with me in the hut when I put before him a gin case bottle of rum nearly full & a tumbler which he finished by himself in less than two hours' time, then went away home but little intoxicated with what he had drank. However from the freedom he had taken with my bottle, on our first acquaintance, I was determined that he should never again have it in his power to enjoy so much from me. Thermometer at 6am 72 at 12 82 & at 7pm 74.

24. A fine day. Got up at 5am & went to the woods through a valley near Diamond Hill, named by myself Wilkinson's, from its being the place where the unfortunate man of this name, who came out with us from England in the suite of the Sandwich Island people with the flattering prospects of improving the state of agriculture throughout these islands, but alas! he found out too soon after landing on their shores that he had to deal with ignorant unfeeling masters to what intent he had come upon for their good, and that instead of employing him to improve their farms with a liberal salary as he at first expected, they have only given him a piece of waste land in this valley to cultivate for himself without any security of being his for no longer period than during their own pleasure. Having to pass through his ground on my way to the wood, I on coming to where he lived enquired of some foreigner whom he had at work clearing the ground for to plant sugar cane if their master was at home, who informed me that he lay ill [in] an old tent (having yet no hut built) a little way from where they were at work. Here I found the miserable object (for he was no other) stretched out on a purser's mattress laid on the ground on a mat, labouring he told me for some day past under the complaint of diarrhea, which had reduced him to a perfect skeleton & the picture of death. During the time I remained with him he often shed tears at his bad fate, & when I recommended that he should apply to Lord Byron for a passage home with us "no" says he, "be my fate what it will, I am resolved never to see England again, I will sooner die where I am in this valley". This farm is situated near the head of the valley, which end[s] in a kind of amphitheatre overhung by high volcanic ridges on each side, thickly covered with candlenut (*Aleurites*) & other trees, that occasion much rain during the night, particularly which will from his damp situation make him to be always sub-

ject to his present complaint. In several parts of this valley I met with great quantities of turmeric, ginger, ava, & one species of *Canna*, growing wild among the grass. The ginger is met with in abundance every where on coming to the woods within half a mile, but afterwards is lost on getting some miles farther. The roots are generally bitter & have not much of the taste of ginger that is cultivated. When in the wood I met with four more different species of *Metrosideroses* not seen before & several other plants. At the head of the valley on the sides of a small river coming from the mountains, are trees of the *Eugenia malaccensis* forming a wood of themselves. On my returning home at 5 o'clock, I passed several hundreds of the natives near where I lived, preparing ground for Mr. Pitt to plant sweet potatoes, by pulling up the grass with their hands, & others using short pointed sticks to loosen the ground ready for planting. They are yet unacquainted with any other description of tools, used by us, to facilitate labour. Thermometer at 5am 70 at 12 83 & at 7pm 75.

25. A fine day. Employed at home among my specimens & seeds. At 3pm called two of the late Tamahamaah's Queens ([Kaumanna] & Pio) on their way to a fish pond near at hand where they generally bath. Both their Majesties on coming up to the door in a cart drawn by several natives, was without ceremony tilted out (just the same as rubbish) being unable of themselves to get out in any other way from their enormous size. They were both dressed in black Canton crepes made in European fashion & had on their heads old coarse straw bonnets but without stockings on their feet, where they had only old slippers. Mr. Forder amused them while they remained, showing them his drawings of plants, which surprised them to see the figure so much resemble the living plants, who they without any ceremony asked to be favoured with the whole that had been shown them, which on being refused they asked for something else, such as pined melia [?] rubber &c. So covetous is the desire of all these people to get every thing they have not themselves, whether they have uses for or not. When they went away the cart which had remained tilted up, both now to get into it, laid hold of the front part standing up together behind, & in this manner the box was raised up with them, with their faces downwards, & their legs (nearly the size of a man's body) hanging over the hind part of the cart, which was too short for their length. Thermometer at 6am 74 at 12 87 & at 7pm 76.

26. During the morning several light showers of rain. Between four & five o'clock am went to the valley above Diamond Hill, on purpose to meet with

Fish Ponds at Honoruru, Oahu. From Maria Callcott's *Voyage of H. M. S.* Blonde *to the Sandwich Islands, in the years 1824-1825. Captain the Right Hon. Lord Byron, commander.* "Engraved from a Drawing made on the spot by Robert Dampier."

Sanders wood, which has bestowed so much wealth on its owner & caused greatly the neglect of cultivation throughout these islands since its discovery, owing to the natives being taken away from their homes for weeks together many miles to procure this article of trade. In the wood at the head of this valley I remained till dusk, having nearly crossed the island & on my return to the first huts out of the wood, where I wished to stay for the night, I found they were all so crowded with natives & otherwise filthy with hogs & dogs inside, that to remain for the night would be very uncomfortable, therefore although nearly 10 miles from home I was obliged to leave much against my will, being anxious to have another day's travelling more to the eastward in the next valley. While I was in the woods I met with many plants not seen before, particularly ferns & [*Metrosiderosis*], the latter formed the greatest number of the largest trees I had seen, excepting the *Acacia* whom the natives' canoes are made from. The Sanders wood I met with grew only to a middling sized tree & was too early in the season for its time of flowering. Bananas grew almost in every ravine wild in the midst of the wood, to a great size with many large bunches of fruit upon them which the native with me cut & roasted, one of which I tasted had a bitter taste, but by no means unpalatable. The fruit of the wild banana is the chief food of the natives when they are cutting Sanders wood & making their canoes light enough for them to be able to drag them home to their huts, which is often the work of several days. I had an opportunity to-day for the first time to see the natives when they were lighting a fire to roast the wild bananas, which they procure by rubbing two dried sticks upon each other that catch fire by friction after rubbing them about a minute. At 9 o'clock got home, after having travelled for several miles in the dark over lava, tarro, ponds & ditches. Thermometer at 5am 72 at 12 87 & at 7pm 76.

27. A fine day, but showery in the morning. Employed till late in the afternoon among my specimens, seeds & skinning birds. Being informed during the day that the ship would leave in a few days for the island of Owhyee, I went to town for to ascertain from His Lordship if he intended to return again to this island, before he left the islands for Otahite, as it would be necessary should he not return to have all my plant boxes on board. Upon seeing him at his house he could only inform me at present that he would sail in a few days. While I was present with him, he introduced me to Mr. Young, lately arrived from Owhyee & whose name with that of his companion Davis, have for a long time been distinguished in the history of these islands, is now fast sinking under the infirmities of old age. In the burial place of a few Euro-

peans underneath a grove of coco-nut trees, near to where I live I read some evenings ago the simple monuments erected at the head of Mr. Davis' grave, having the following lines upon it:

The remains
of
Mr. Isaac Davis
who died at this
Island, April 1810
aged 52 years.

Mr. Davis at his decease possessed many lands or farms throughout the different islands, given him by Tamahamaah for his faithful services, which afterwards became the property of his wife & three daughters, until lately when several have been taken from them by some of the covetous chiefs now in power, much to their discredit. In the evening His Lordship with several of the officers on board met at the young King's hut, where was performed the Magic Lantern, the same as when crossing the line before our coming to Rio, for the purpose of amusing His Majesty & sister, but who from the religious fanaticism of the American Methodists, was by them both prevented from having the pleasure of being present. These missionaries, many of them who are but illiterate mechanics possess what they please over the credulity of the natives, & have already carried their system of religion too far ever to be upheld. Thermometer at 6am 75 at 12 85 & at 7pm 77.

28. A fine morning. Met by appointment Mr. Charlton at 6am at Mr. Pitt's hut for the purpose of seeing the plants brought out from England transplanted, but on my arrival found to my surprise that the natives had already finished planting out one of the boxes of tropical plants, in which I had several *Orchideceas* from St. Catherines doing well, although my *Orchideceas* intended for Mr. Lindley was now partly destroyed. Yet my feelings at the moment were nothing compared to that of seeing them tear up my others by the stems, as a gardener would cabbage stumps to be thrown away on a dung-heap. Mr. Charlton whom I expected to be present was not yet come. I therefore went to Pitt to request that he would order the natives to give up doing anything to the plants, till I returned with the consul, who only laughed at my [eager] confused state at the time, & found when I had returned with Mr. Charlton that they had now the assistance of the well known Spaniard

(Marin) mentioned in the history of these islands, particularly by the celebrated navigator Kotzebue, in whom the government put but too much confidence in, so favourable to his own interest, generally at their expense. This self-conceited Spaniard kept in pulling up the plants by their stems in my presence, then giving them afterwards to the natives, with their fibers deprived of earth for them to plant as they chose without considering any distance or regularity for them to grow upwards with sufficient room afterwards. When I remonstrated with him for tearing up the plants by the stems, he had the impertinence to tell me that they would do especially as well, upon which I requested he would not touch a single plant, but allow me to finish the rest according to my way of thinking. Mr. Charlton having noticed my feelings at the time from the manner I expressed myself, took me a short distance sideway, & recommended that I would better leave this fellow himself to do with the plants as he thought proper, to which I consented on condition that he (Mr. Charlton) would explain to strangers touching here, that I had no hand in their transplanting or in choosing the situation which if kept ignorant from them, they would not hesitate to censure me to the world, when they saw in what careless manner they had been planted. Being promised such I then took up my St. Catherine *Orchideae* & replanted them again in the same box where they had been taken from, & next proceeded to take a list of what plants had been introduced alive, but to my surprise found that the natives had stolen all the leaden numbers attached to each plant, as well as the iron hooks which kept the box lids open when the plants required air on the passage. At 10 o'clock returned home to my hut & afterwards employed myself shifting specimens & making up duplicates of seeds to go home by one of Mr. Charlton's vessels bound to the Port of Valparaiso. Thermometer at 5am 75 at 12 87 & at 7pm 76.

29. (Sunday) A fine day. Having promised yesterday to meet our purser this morning at 6 o'clock at the house of the harbour pilot (Mr. Adams) for the purpose of accompanying them to the opposite side of the island to see the steep cliffs where Mr. A. Bloxam and myself had on the 13th some difficulty in passing. I therefore with Mantle got up at day-light & left shortly after to meet them, taking with us our gun & specimen box. Adams amused us much throughout the day relating several curious anecdotes of the late Tamahamaah & his Queens, particularly about Kaumanna. He informed us that he had been on the islands since the year 1809 during which time they have employed him various ways, sometimes to China with Sanders wood & at

others to the north west coast of America with cargos of salt. Where on his voyage the last time he brought home a couple of deer with a view of their thriving in the Sandwich Islands, but they had not long been suffered to go at large in Hana-rura valley, when Pitt a short time after happened to get unwell & fancied that the flesh of the deer would do him good, as Adams had when he brought them spoken highly of their flesh being good to eat, in consequence of which he gave orders that one of them should first be killed for him to taste, which he afterwards found so much to his liking that he again ordered the other to be killed shortly after. This ended the life of both poor Adams' deer & the good which might be expected to arise from the introduction of so useful an animal. On our way through the valley of Hana-rura or Nuana oftentimes called by the natives, we was shown the spot where the king of Woahoo's (Tereaboo) head general was killed, after the invasion of Tamahamaah from Owhyee to conquer this island. The circumstance related to us was nearly as follows, that when Tamahamaah with Young & Davies, & the rest of his army, had landed from their small fleet in the harbour without opposition from Tereaboo, who had collected his forces above the town in Hana-rura valley where Tamahamaah could not have wished for a better situation more favourable to his purpose, being over hung by ridges on each side, left unoccupied by the enemy, where he without any opposition placed a certain number of his men, when himself with Young & Davis & the greater part of his army took the center of the valley, having with them only one small swivel and a few firearms, the rest being armed with spears and clubs. Mr. Davis who had the swivel, somewhat singular to relate, on his firing the first that killed Tereaboo's head general before scarcely any engagement had commenced, which as is customary with all the natives of these islands, they instantly got into the utmost confusion & retreated when they found their leader had fallen. They were then afterwards pursued by Tamahamaah across the island, where on coming to the steep cliff already mentioned they threw themselves over the precipice & were thousands of them found lifeless at [the] bottom when their pursuers came up. In this manner by the help of Young & Davis without hardly any firearms, did Tamahamaah make the easy conquest of this important island which at present may be considered the first of all the Sandwich Islands on account of its good harbour. The King fled to the mountains, being convinced the custom of putting the vanquished to death would be practiced upon [him]. "I must die["] said he to one of his friends ["]for I will not let him enjoy this triumph I will sacrifice myself to the gods." His corpse was afterwards found in a cave in the mountains. At 5pm on our return near the town parted with Adams & the purser & soon after got home to skin some of the birds I had shot. One of these was particularly handsome, being all red & only met with sucking the blossoms of the *Metrosideroses*. A little before dusk called Mr. & Mrs. Charlton & sister, also several of the gentlemen from on board. Thermometer at 5am 76, at 72 86 & at 7pm 76.

30. A fine day, with at times some light showers of rain. Employed at home among my specimens & seeds. In the evening called Lord Byron & the surgeon on their way riding out towards Diamond Hill, with whom I had a good deal of conversation respecting the Spaniard who so ill treated my plants brought out from England. Thermometer at 6am 76 at 12 86 & at 7pm 76.

31. A fine day. At 10am went to town for the purpose of removing my empty plant boxes from Pitt's to the carpenters to be repaired. Saw first on board his own vessel Mr. Charlton who came with me to Pitt's for to request him to order some of his people to take my boxes to the carpenters, which was promised, yet after waiting for upwards of an hour with him in his hut, no one appeared to do this for me, when I then begged of Mr. Charlton to go with me to procure assistance for which I would pay the expense sooner than remain here losing more time depending on Pitt. Having met on our way with Adams the pilot to whom we told what we wanted, who very good naturedly went to a friend & borrowed a cart, then got his own people & came with them to assist me, when we soon succeeded in removing the whole. Before I left Pitt's I examined the plants who still remained without any artificial shade being put over them to protect them from the hot rays of the sun so common at this season of the year & found but only one grape vine among the whole that had any leaves upon them. The entire neglect of these people towards so many useful plants introduced for their benefit at so much trouble & anxiety for preserving them alive during a passage of nearly eight months & a distance of upwards of 15000 miles through various climates at considerable expense to the Horticultural Society, easily shows how their ignorance yet in the arts of agriculture. In fact the labour & expense of these plants may appear in a manner to be thrown away upon them a century too soon. Still if it is to be hoped however that some out of so great a number as 138 will not altogether be lost in the country. The Spaniard (Marin) whom I have already had occasion to mention, has had the grape vine in his possession since 1814 from California at which place he had been with Adams as linguist in one of the Sandwich Islands vessels to bring over cattle, horses, & mules &c., which

on his return he planted the grape vine in his own ground near the town of Hana-rura & got the King to perform the charm of taboo to prevent any being stolen which till the present time still continues to be in force. So selfish is the disposition of this man that although he has succeeded to cultivate the grape vine with much success & within the last two years to make more than three casks of wine from the fruit, that sooner than he would part with slips to his neighbours he actually burned the prunings every year rather than oblige those who wished for to cultivate them, that he might enjoy the monopoly arising from them himself, which will now soon be out of his power from the liberality of Mr. Charlton who has given away to different individuals several fine grape plants brought with him from the Port of Valparaiso, besides an assortment of vegetable seeds from England presented to him by the Horticultural Society. Mr. Charlton who possesses liberal principals will soon reside at Woahoo constantly with his family, where he intends to grow coffee & cotton, which has yet been neglected. It may therefore be expected in a climate so favourable to his views, that a change of improvement will soon follow, not even in cotton & coffee, but in coco & sugar, where the soil & climate is equally favourable for it, to be brought to the same perfection with any in the world. Thermometer at 6am 76 at 12 87 & at 7pm 77.

1 June. A fine day, but somewhat showery in the forenoon. At 10am went to town with the few living plants collected for England to get them planted in one of the empty boxes promised by the carpenter yesterday to be repaired. When returning home saw Queen Kaumanna at a short distance from the road having several natives with her dragging her four wheeled cart to the top of the small hill, which was afterwards pushed off to go to the bottom by itself alone with her in it. This ludicrous sort of amusement was always accompanied by the natives with much shouting noise while her Majesty & cart kept going to the bottom. At 2pm called the Captain's clerk to inform me that he had just come on shore to write dispatches for His Lordship to go home by the *Active* bound to Valparaiso in a few days. I therefore in the evening went again to town to try His Lordship once more to allow me the permission of altering his former request of directing my packets of seeds to Mr. Croker instead of that to Mr. Banning whom through the latter Mr. Charlton like Mr. Nugent at Valparaiso kindly offered to forward any packets of seeds or letters to the Society. On my coming to town I met with His Lordship & the surgeon on their way riding out to call upon poor Wilkinson, & who upon communicating to him what I wanted, he after some hesitation granted my

request, observing at the same time that he could not see why I should not be at liberty to direct to the one the same as the other. Thermometer at 6am 76 at 12 87 & at 7pm 76.

2. A fine day, but somewhat showery in the forenoon. Employed at home sorting duplicates of seeds to be sent home by the *Active*. In the afternoon called lofty Queen Kaumanna & Pio with several other female attendants, besides the natives dragging their carts. Both were very inquisitive to knowing if my friend Mr. Forder & myself had been married men, who on their being informed we was, they next enquired what was the number of our family, to which my friend who kept up the conversation with them, said that he had six, but that I had none, at which they remarked I must only have had but one wife. They then after some more trifling questions left us for to bath in their usual fish pond at a little distance from our hut. We was again a little before dusk visited by the young King on horse back accompanied by a little boy brought out from England by Boki. His Majesty was clothed in a short blue jacket & pantaloons of the same colour, but was without a hat, waistcoat, shoes or stockings. We afterwards noticed the young Princess his sister carried home from the bathing place on a native boy's back at some distance on the road before the old Queens dragged after her in the carts. Thermometer at 6am 77 at 12 87 & at 7pm 76.

3. A fine day, but as yesterday showery during the forenoon. As the ship sails for Owhyee on the 7th, I am requested to remove my traps again this evening on board. In the forenoon shifted my specimens & in the afternoon packed up the same & got everything on board by dusk. Thermometer at 6am 74 at 12 85 & at 7pm 76.

4. A fine day. Busily employed during the forenoon getting my things stowed away in the midst of noise common to a man-o'-war. At 5pm one of the gentlemen hooked a shark alongside, which measured 11 feet in length & five in circumference & who after being cut open, was found in the stomach a large hook & chain carried away the day before besides a bullock's foot, part of a pig's head & a variety of other bones, which had been thrown over the ship's side, where he had been noticed for several days past. In the afternoon a quantity of luggage & provisions was sent on board by the Sandwich people for their sea stock while on their passage with us to Owhyee. Thermometer at 6am 75 at 12 77 & at 7pm 75.

5. (Sunday) A fine day. At 10 church service performed as usual. Employed all day writing out my accounts, as well as letters for Mr. Sabine. Thermometer at 6am 76 at 12 78 & at 7pm 76.

6. A fine day, went on shore at 10 o'clock with my packet of seeds, letters &c. to be left with Mr. Charlton, when afterwards I got Lord Byron to endorse me a bill for £20, which I sold at the rate of 5 [shillings?] per dollar. While in my travels through the town saw a large pig roasted with hot stones in the customary way with the South Sea Islanders mentioned already by Captain Cook & others & shortly after noticed a young woman suckling several puppies, lashed up [with] a piece of tapa cloth wrapped round her shoulders & breasts, as she kept walking along the streets. The custom of suckling dogs & pigs &c. is common to the natives of the Sandwich islands & is held by them in estimation, little inferior to their own offspring, which on my journey to the woods in search of plants often afforded me the opportunity of being an eye witness to, while I had occasion to enter the huts, sometimes from curiosity & at others for a drink of water, I not only observed the suckling of dogs & pigs, but also even feeding them with the poi made from tarro root, which they chiefly subsist upon themselves & feed these animals with it the same as a mother would do her own child, which often gave me such disgust that I was obliged to leave their hut immediately. Their dogs are in general, useless for any thing else but being eat, & seldom ever heard to bark. In size they are small, with long bodies, & ears, sharp piked noses & short feet & mostly of a black colour, but are very often seen covered with mange to that degree, so as to be almost deprived of hair, yet the natives [feed] them in this state & preserve all their offspring, which they carry in their arms oftener than do their children & feel more offended if a stranger passing their huts was to hurt them, than they would do for the other. At 2 o'clock a general assembly of all the chiefs with Lord Byron & others took place at Pitt's house for taking into consideration fit laws to be established throughout the different islands. When Boki stood up for half an hour and made a surprising speech, which astonished all that were present, particularly some of his own countrymen who said it must be some other person in Boki's skin. He commented at great length upon what he had seen in England compared with that of his own country, & strongly recommended them to establish laws as well as religion on the same [principles] of that in England. He said he was sorry to observe that their present religion restricted them from lighting fires on Sundays to cook their meals or to bath themselves contrary to the custom

of England, but for his own part he was determined to have a fire in his house when he wanted one, or to bath himself as he had seen them do in England. After several discussions among the parties present, it was then resolved that the former laws of Tamahamaah should be again put in force. About three o'clock the marines came on shore to go through several military movements for the gratification of Pitt and others, but unfortunately many of them got intoxicated on their way through the town to the ground, which to the disappointment of His Lordship several of them were unfit to go through their respective movements. There was present at the review most of the Royal family & chiefs besides the Europeans residing in the place. Among those that were present I noticed Queen Kaumanna in her cart with the young King half naked sitting above her while she lay in the cart, who upon every time the marines fired squatted himself, with his face downward on her Majesty's back. While at the review I met with Mr. Wilkinson brought from England by Boki, who has got now much better of his complaint & is very anxious to get His Lordship before he leaves for Owhyee to persuade Pitt, to promise him that the small spot of ground given him when he has improved it, may not be afterwards taken from him, but unfortunately for him like the rest of the islands, he informed me that they will not present through any recommendation of his friends promise him any farther possession to remain, but until the ground may be wanted by themselves. Having to procure some things in the morning before the ship got under weigh, I remained at the inn of Novara in company with the assistant surgeon. Thermometer at 6am 76 at 12 79 & at 7pm 76.

7. A fine day. At 10am called upon His Lordship to ascertain whether I was to take my empty plant boxes on board, who informed me there was no occasion as he intended to touch here again. At 12 Queen Kaumanna & Pio, with Mr. Young & several chiefs & natives embarked to go with us to Heddo Bay situated on the north east side of Owhyee. At one pm took leave of Mr. Charlton, with hopes of seeing him soon again when we touched on our passage homeward at Otahite, where he had promised to render me every assistance while the ship remained. Returned on board a little before 2 o'clock & got under weigh at three with a fresh breeze from E.N.E. At dusk the wind still continued fresh & cloudy. Among the Queen's attendants, are an old cunning waggish fellow, named Jack Bligh a native of Otahite, who spoke a little English, & had he informed us been with Captain Bligh in the *Bounty* at the time the unfortunate mutiny took place, whom he described in much the

Lord Byron's House at Byron Bay. From Maria Callcott's *Voyage of H. M. S.* Blonde *to the Sandwich Islands, in the years 1824-1825. Captain the Right Hon. Lord Byron, commander.* "Engraved from a Drawing made on the spot by Robert Dampier."

same way as given by Bligh himself. We have on board besides those already mentioned two of the missionaries & one of their wives come on board for the good of her health, also Sir Joseph Banks as an interpreter to Lord Byron. Thermometer at 6am 76 at 12 78 & at 7pm 76.

8. A fine morning with fresh breezes from N.E. at 10am articles of war was read to the ship's company. Then after had 10 of them punished, some for drunkenness, & others for disobedience of orders. At noon the island of Mowee in sight. At 6pm the wind shifted to the north, moderate & cloudy. Thermometer at 6pm 73 at 12 75 & at 7pm 75.

9. A fine morning with fresh breezes from W. by South. At 4pm the wind ceased to a calm, being now near in shore off the land of Mowee, a boat with Joe Banks was sent on shore to some huts near the sea beach to procure fish, who after returned at five o'clock with a large quantity of fresh & salted, which the natives immediately after getting them on board, began to devour the fresh in a ravenous manner raw, with some poi & tarro, not even wasting their gills or entrails, but considered them the most delicate parts of the fish, & what was generally eat by the Queens and chiefs. Among those eating fish in this disgusting manner, we noticed to our surprise Manaware who we had brought out from England seated in the midst of them with his shirt sleeves rolled up to his shoulders, eating & tearing to pieces with his teeth the gills & entrails having his mouth, hands, & fingers besmeared with blood, from his not using a knife & fork, a thing yet never adopted by the Sandwich islanders. I was so much disgusted on seeing this fellow, who of late had been accustomed of seeing a different way of living, that I could not help be without telling him that he ought to have shown his countrymen how he had seen us eat our meals, to which he replied somewhat offended, that he & all the people of his country liked fish in this way best, observing at the same time, he see plenty of poor people in England, but we see none here that they got plenty of poi, tarro, & fish, & no want for any, like many poor man at home. Such was the answer I received from a person who had been in England with their King, & who on the passage like the others of his countrymen, to their own country was fed by Lord Byron in a style not inferior to what he usually had himself. At dusk still & fine weather. Thermometer at 6am 74 at 12 76 & at 7pm 75.

10. A fine morning with fresh breezes from N.E. 7am saw Mouna Kaah far above the clouds in places covered with snow, which had a grand appearance.

At noon & dusk still fresh breezes, with after dark some light rain. Thermometer at 6am 73 at 12 75 & at 7pm 75.

11. A fine morning with strong breezes from East. At 8am Mouna Roa with its round flat top in sight bearing South & Mouna Kaah N.E. as also the high land of Mowee about north. During the day several showers of rain. At 7pm fresh breezes & cloudy. Thermometer at 6am 73 at 12 76 & at 7pm 75.

12. (Sunday) A fine morning with strong breezes from E.N.E. & cloudy. At 10 church service performed when the Queens & several chiefs were present, as also the missionaries. At 3:40 shortened sail & came to anchor in 6 fathoms. When we had come to anchor I spoke to Lord Byron to ask Queen Kaumanna for natives to go with me to Mouna Kaah, where I intended to undertake the journey as soon as possible & thereafter my return, to make the best of my time in the neighbourhood while the ship remained. His Lordship through one of the missionaries who acted for him as interpreter communicated to her Majesty my intended journey to the snowy mountain & the necessity of having 7 or 8 of the natives to accompany me, which after some hesitation to consider, as is customary with her, she at last gave her consent to have what number I wanted when I went. Having now so far settled things to my satisfaction, I had yet still another favour to ask from her of being provided with a hut somewhere on shore to remove my traps to, tomorrow, where Mr. Forder will live till my return on purpose to receive, & dry what plants I may find necessary to send home while on my journey. To this request I was desired to be informed that she at present did not know of one, but when she went on shore herself, she would enquire of the chiefs if one could be had for me. Thermometer at 6am 73 at 12 77 & at 7pm 77.

13. A fine day. Went on shore shortly after day-light to find out the huts of the only two foreigners at this place besides the missionaries, to procure one of them as guide to Mouna Kaah. I had not long been on shore when I accidentally met with one of the missionaries (Mr. Goodrich) brought with us from the island of Woahoo, who informed me that both the persons I was in search of, had left here a fortnight ago to kill wild cattle near Mouna Kaah & would not return perhaps for some weeks, but before I should be disappointed for want of a guide, he would willingly accompany me, being once there before himself. This kind offer completed my wishes & had only now to arrange between us what things was requisite to carry on our journey, & whether it

was best to go part of the way by water in a canoe to get nearer the foot of the mountain & save a journey of 30 miles by land along the coast intersected with numerous deep ravines & large rivers, where after travelling we could not gain the summit of the mountain by any other path than that where we could land by water. It was therefore thought advisable to go by water & on our return to come home by land, as we have no difficulty to procure a canoe if I made again application to Kaumanna who had plenty at her command. On my return on board to breakfast I shortly after informed His Lordship what arrangement I had made & begged the favour that he would apply to her Majesty for a canoe & extra natives to man it, besides those already promised me, which in his usual pleasant manner he promised when he found her in such humor, as likely not to refuse him, being at present rather sulky from accounts received of some persons on shore having acted wrong in her absence. I had not long been on board when my intentions of going alone to Mouna Kaah was known among the officers, many of whom wished to accompany me if I waited for a few days, others said they would wait & go with the surveyor when he arrived in the schooner from Woahoo, where he was left behind to finish the survey of the Pearl River. Of those that were most anxious to go with me, was Mr. Talbot 4th Lieutenant & Mr. Wilson purser, if they had His Lordship's permission, which they requested me to get for them in preference to themselves in case of their being refused, which on seeing him soon after I mentioned their wish to be of my party if he gave them permission, to which he said he was perfectly agreeable & at the same time acquainted me that Kaumanna had promised me a canoe & natives tomorrow. At noon went on shore to look among the native huts to choose one that I thought would answer me to live in, which was promised to be given me by Manaware brought out from England. After getting on shore I first went to Mr. Goodrich, who went with me to most of the huts round the Bay, all pleasantly situated under the shade of bread fruit trees, which in places form woods of themselves & grow to a great height, producing plenty of fruit, but they possess but little variety & are generally of the small kind. There is also here plenty of both kinds of the rose apple & of coco-nut trees, some of which are very great height & show the appearance of old age. Heddo Bay & its neighbourhood is a very pleasant spot, on account of its woody appearance & plentiful supply of water & is far superior in point of scenery to any part which we had seen, either at Mowee or Woahoo, but like there is little cultivated with only small patches round the native huts, of bananas, tarro, tapa &c. At the west end of the bay is a considerable sized river with several water falls, convenient to water vessels laying in the harbour, & opposite the center of the bay are another subterranean one which is dammed up a little distance from the sea & forms a large fresh water fish pond, taboo'd for the use of the King & the chiefs, & at the east end not far from the shore, & near a reef which run partly across the Bay there is a small island thickly covered with low coco-nut trees, which gives it the name of Coco-nut island. The whole of this side (north) of Owhyee which is divided into two districts belongs to Kaumanna & Pio their place of residence when at Heddo to receive the rents is near the east side of the Bay, which is no more than two huts or rather sheds, one of which is given to His Lordship to reside in while he remains. At dusk returned on board, when shortly after I learned that the canoe & natives could not be ready for me tomorrow, but would on the following morning. In the evening on deck I had some accounts from Mr. Young of Dr. Menzies' journey to Mouna Roa next in height to Mouna Kaah to where I am going. Mr. Young's opinion of the height of the latter mountain, being much higher than the former goes no farther in proof of their difference than that during the 36 years he had been on the island he never noticed Mouna Raah free from snow, but has always seen the other without any in summer, which on that account he had considered it to be inferior height to Mouna Kaah. Thermometer at 6am 74 at 12 76 [&] at 7pm 75.

14. A fine day. At 10am went on shore with my traps, salt provisions &c. taking with me Mantle & another lad named Trounce who are both allowed me from the ship while I have occasion for them to assist. When I landed with my things I found to my surprise that the hut promised me by Manaware was now refused & only offered part of another at the opposite side of the bay inhabited by a chief. Being thus disappointed for want of a hut to secure my provisions & other things now landed on the beach surrounded by crowds of natives, who if they had the opportunity would not hesitate to make free with what they could lay hold of, I therefore begged permission of His Lordship to allow me to put part in a corner of his house to which he kindly informed me that it made no difference to him leaving any of my things in his hut till I returned & that he should order care to be taken of them while they remained, but if I wanted a place for Mr. Forder & them in preference to his hut, the tent near at hand put up for his servants was at our service, which I accepted. Mr. Young & two of the missionaries being at the time present with him I enquired if they thought Kaumanna would disappoint me with the canoe & natives tomorrow at day-light as she had already done & whether it

would not be advisable to remind her of it again, while so many were present whom she could understand, which they agreed to be of my opinion & willingly offered to go with me to her hut, when we found her as usual lying on the floor with her face downwards & several natives round her brushing the flies from her body. On my informing her that I wished to know if I could depend on the canoe & natives to come for me at day-light tomorrow morning she hesitated in giving an answer till she had satisfied herself in taking a survey of me from head to foot, then said when she saw one of the chiefs she would let me know. On her giving me this unsatisfactory answer, I requested Mr. Young who had more influence over her than the missionaries had, to tell her that if she could not let me have the canoe [and] people conveniently for nothing that I was willing to satisfy her or the chiefs with what money [they] wanted if they promised to let me have one and natives tomorrow, being determined to go on my journey whether they gave me them or not. When Mr. Young acquainted her with what I had said I was glad to find it had the desired effect, for she instantly sent a canoe across the bay to the head chief, who in a short time afterwards soon came to her, when they settled the business between them, that I was to have the canoe & people at the hour I wanted without paying any money whatsoever. Having at last succeeded with her Majesty to grant my request, through the help of Young & the missionaries, I sent a message on board to acquaint Mr. Talbot & Wilson to be provided with every thing they thought necessary wanted by themselves on the journey to-night, so as to prevent delay in the morning when I came alongside for them with the canoe. In the afternoon the lads boiled two pieces of salt beef, the one for the dinner & the other to take with us in the morning with some more pieces uncooked, which with many others things likely to be wanted, took up my time till after dark in getting them packed up. At 4pm His Lordship asked me to dine with him, where I met present the surgeon, chaplain, & painter, who are his general constant companions while on shore. At sun-set Mr. Forder joined me from on board to remain on shore. When both afterward took our lodging for the night in the tent with two of His Lordship's servants. Thermometer at 6am 74 at 12 79 & at 7pm 76.

15. A fine day, but in the night several heavy showers of rain, which came pouring through the old tent & made us very uncomfortable during the night. At day-light arrived Mr. Goodrich with the double canoe & natives, when we soon after began to embark our provisions & luggage intended for our journey, but on account of some of the natives being still absent getting

their own provisions of poi & fish, we was detained till nearly 6 o'clock before getting alongside the ship for Mr. Talbot & Wilson, whom we both found ready waiting for us & soon after took into our canoe, then left the ship (now in number 17: eleven natives & six ourselves) with the well wishes of those on board for our success & safe return, to Lapahoi a distance of 30 miles on the west side of the island, having a favourable light breezes from easterly, which kept freshening stronger every hour till our landing at 11 o'clock, in a narrow creek on each side full of rocks open to a high surf, that is generally here on the coast, which at all times excepting early in the morning make landing very difficult & dangerous, as we ourselves experienced in having the greatest difficulty to manage our canoe from being dashed on shore upon the rocks, owing to the surf washing over us every minute, which kept filling the canoe with water so fast, as to render our efforts in baling it out useless. When we had partly shifted ourselves with dry clothes, & finished rubbing our forearms free from the salt water, we examined the rest of our luggage to ascertain if anything had been washed away by the surf while in our confusion landing & found to our disappointment that 40 lbs of salt meat which I had for my share of provisions was missing, but nothing else. Lapahoi is a small stoney flat with a few huts, & sweet potatoes & tarro patches scattered over it, situated at the extremity of a deep ravine, the declivities of which on each side are nearly 500 feet in height & extend to the sea beach, terminating in a rocky precipice. The coast all the way to Lapahoi we noticed was intersected by numerous deep ravines, many of whom have large rivers forming beautiful waterfalls over the outward cliffs falling at once into the ocean, the angry surf of which broke a long way up, upon the rocks underneath. In most places along the upper part of the abrupt declivity, grew plentiful a species of *Pandanus* used common by the natives to make mats for the floors of their huts, which here form of themselves almost everywhere thick plantations that give the coast a very pleasant appearance at seeing their green bushy tops hanging pendant over the rocks where underneath in many places small subterranean streams of water fall down at no great distance from each other. This species of [*Pandanus*] is nowhere so plentiful in the Sandwich islands, as it is at Owhyee, in the others it is cultivated frequently for its leaves to make mats & pillows for the natives to sleep upon. The tea tree is also here plentiful in the ravines everywhere along the coast. By noon we had finished taking some refreshments & getting our baggage equally divided into loads for the natives to carry when we then began to proceed on our journey, leaving behind six with orders to remain for four days, with the canoe in case we might return by that time & go home by water, taking with us the other five, which including ourselves six made in all eleven in number. Upon gaining the summit above the huts of Lapahoi, we here remained a few minutes to draw breath, looking all the while upon the high surf breaking over the rocks underneath where we had, but lately so narrowly escaped landing with our lives. On resuming our journey we had not got far when it always became more interesting every step as we kept advancing in a narrow path leading towards the woods above, yet about four miles distance which as we kept approaching, the few native huts seen on each side were fast disappearing. The whole face of the country from the sea coast to about six miles in the interior to the edge of the wood produced various fine prospects, which reminded us much of home & if cultivated appeared to be capable of producing an equal return of crops to any land of a similar climate, is yet suffered to remain as nature had formed it, without even pastured by any kind of livestock, covered with long grass & a number of short stumpy tree ferns belonging to the *Cyathea* tribe, the roots of which afford food for the herds of swine that are generally seen about the thinly scattered huts of the natives, who cultivate little patches around them filled with sweet potatoes, tarro, sugar canes & a few bananas, sufficient to supply the wants of the naked beings that inhabit them, which often amount from three to four generations, huddled close together at nights in one hut like so many dumb animals, without at times sufficient shelter over them to protect them from the cold heavy dew that commonly fall here invariably at nights. Between three & four in the afternoon we reached the outskirts of the wood at a few native huts, having crossed on our way three narrow deep ravines thickly covered with wood chiefly *Metrosideroses*, *Aleurites* & a species of *Rhus*, but without water excepting during heavy rains. Our guide (Mr. Goodrich) now recommended us to take up our quarters at these huts for the night, although it was early in the afternoon, as we had come to the last huts to be met with inhabited on our way to the mountain where we had any chance to procure what we wanted to eat, to take with us in the morning, which on account of the provisions of mine lost we could not do without, unless we met by chance with the two Europeans that were killing wild cattle in the direction of our journey. When we were about to enter one of the largest huts to see if it was free from filth to stay at for the night, Mr. Goodrich was friendly accosted by a smiling active young woman whom he knew was the wife of one of the men killing cattle, who informed us she had only left them yesterday morning & that they had shot two bullocks the day before she left. This information proved to be satisfactory, that we now only wanted to procure from the natives sufficient to answer our wants for the night, which upon taking possession of the cleanest part of the hut for our accommodation without leave or ceremony, as is customary with these people themselves, Mr. Goodrich & myself went among the rest of the huts to purchase a young pig or a couple of fowls, but after several offers of both money & looking glasses we only succeeded in getting a couple of fowls, owing to the price put upon their pigs, being nearly triple what they were worth. On our return to get these cooked we found Mr. Talbot & Wilson who we had left, none highly amused seated in the middle of a large crowd of natives collected from all quarters eagerly reviewing them with surprise. While supper was getting prepared I took the opportunity of going to the wood, which was only a short distance off to look for plants, where I remained till dusk, having met with several species of ferns not seen before, besides a few other plants. Being yet today scarcely beyond the outskirts of the wood the trees as far as I had been were but a moderate size & generally consisted of *Metrosideroses* & *Aleurites*, with abundance of ferns growing underneath their shade. Among the different species of *Metrosideroses* in variety of colours of the flowers as well as foliage already met with before at the island of Woahoo, there still appears in this part many which will add to their number, one particularly with straw coloured flowers & which underneath the leaves met with this evening but rather sparingly adds to my former number. When I got home to the hut at dusk my three fellow travellers was sitting on a mat, each having part of a fowl in one hand & a clasp knife in the other busy eating in the presence of a number of natives, two of whom held in their hands a light made from the kernels of the Ku Kue or candlenut (*Aleurites*) severals in number passed through on a splinter of the bamboo cane, which gave a greater light than two or three common sized candles. About nine o'clock we retired to sleep in a corner of the hut on a clean mat brought with us on purpose, which every other part was occupied by natives, men, women & children lying mixed together, nay even their scabby disgusting dogs shared part of their bed, who they were seen to take fondly in their arms to lie beside them. The scene around us was so interesting particularly to Mr. Talbot & Wilson & myself that we kept on talking & laughing for hours in the dark till at last overcome by sleep. Thermometer at 6am 76 at 12 80 & at 7pm 71.

16. A fine morning, but somewhat foggy and during the night rather cold from heavy fall of dew. Got up at day-light & took the temperature of the air which stood 64 & at five o'clock every person was ready to proceed on their

journey regardless of the heavy dew that still continued on the grass and bushes, which before we had travelled far soon wet us through above our knees. Having entered the wood about a mile to the edge of a small ravine by a narrow path where on each side, grew in places a number of strong healthy banana trees without cultivation, many of whom had large bunches of fruit. At this ravine our guide informed us, was the place where Mr. Young & his comrade Davis fought their first battle in the service of Tamahamaah & defeated upwards of 10000 of the enemy with only 300 before their leaders came up to their assistance with the main body of the army. The description related to us of this engagement, was that when King Tamahamaah had conquered the south side of Owhyee he soon after with his army marched round to the opposite side of the island by the eastward taking with him Young & Davis for the first time to whom he gave command of the chief part of his army [ ] the chief of the north side in this part named Heddo was prepared to meet him to defend his proportion of the island being subject to his power. But who on seeing the superior force of Tamahamaah kept retreating to the westward till overtaken by Young & Davis nearly a day journey in advance from the main body of the army. The attack took place early in the afternoon from the opposite sides of the ravine in the wood, when after several hours engagement it was decided in favour of Young & Davis, who alone both had firearms that killed the enemy in vast numbers from the crowded manner in which they stood to oppose them, being unacquainted with the destructive effects of firearms. This battle gained to Tamahamaah the conquest of Owhyee. At 9 o'clock we halted to have some refreshment, having travelled about four miles through the wood, here I [had] an opportunity to ramble a little way out of the path while my companions sat on an old tree that lay across the footpath covered with moss & ferns. Some of the trees had now as we passed on, began to appear every step more lofty particularly a species of *Acacia* used by the natives for canoes. Ferns of all kinds & sizes covered the ground underneath the trees & even a good many grew as parasites upon their trunks. A noble species of *Cyathea* seen equally numerous with the rest often attained the height of 25 feet & nearly 12 in circumference. *Metrosideroses* with red bunchy flowering tops, having round them numerous red birds sucking their blossoms, attained now here a much greater size & height than any I observed on the island of Woahoo. *Beslerias* of various coloured flowers, which severals were of a climbing nature, with a numerous tribe of *Psychotrias* both shrubby & succulent, as also many *Lobelias* & other plants added in their variety to enliven our journey truly pleasant, notwithstanding the difficulties often met with from the number of large trees laying across the path every other short distance which we had in general to scramble over, sometimes attended with falls, being wet & slippery from the almost constant rains that fall at nights. By 10 we travelled about nine miles when we again halted for a short time to fill our calabashes with water, having got to the last place, where water could be found till our return from the top of Moana Kaah, in this place I again took the temperature of the air, which was risen to 69. We after leaving here resumed our journey over trunks of trees as before with much the same kind of road, but raising much more in ascent towards our getting out of the wood, where raspberries, strawberries & *Vacciniums* now began to show themselves abundantly on every side covered with fruit which we all eagerly eat of heartily & found much relief from by quenching our thirst. The raspberries were in general very large & in shape flat at both ends, but were round in the middle & although growing wild not impleasant in flavour. The strawberries were small & like the other not of a bad flavour. The *Vacciniums* had an acid taste, yet not altogether impleasant to eat, many of the plants on which they grew attained the height of fifteen feet. During the last three miles of our journey from the great quantities of fruit seen everywhere around us, we at times fancied ourselves, as if we were wandering through a garden neglected to a state of wilderness. Having at one pm succeeded in travelling through the wood, a distance of twelve miles, & upwards of twelve thousand feet above the level of the sea, here we soon after found out the temporary hut of the two Europeans at this place for some weeks killing wild cattle that had been introduced by Captain Vancouver from the northwest coast of America & since suffered to remain unmolested for upwards of twenty years. But since the death of King Tamahamaah the government have frequently killed & salted many for the supply of their small fleet. In the hut we found both at home asleep dressed in the costume of the country & about twenty natives, men, women & children outside some lying asleep, others roasting pieces of flesh stuck on the point of a stick fixed in the ground in a leaning posture over the fire. Both being well known to any guide, they were soon made acquainted with the object of our visit & readily offered to supply us with what beef we wanted, which at this moment was to us of great consequence from our having lost most of our salt provisions taken with us from on board the ship. While the natives were cooking some of the beef for our refreshment, we learned from these two half naked foreigners who could speak but indifferent English, although one a Welshman the other a Prussian blacksmith, who were both for some time once in the English

Navy, that they had succeeded during their short stay in shooting several cattle, but not without some difficulty although often seen in droves to the amount of more than twenty, being always sensible of any person approaching them & if unsuccessful in killing them with the first shot, a place of retreat for their own safety, is instantly required, as they invariably pursue their destroyers with a kind of furious madness, while they appear in sight. Besides this we was further informed that they had two days since killed an old black bull, which they thought was one of the original brought by Vancouver, from part of the right ear being cut off, a mark which they had been told was put upon them when landed upwards of thirty years ago. They have now increased since that time to some hundreds, yet it may appear somewhat curious that they have never been seen more than a few miles in downwards in the woods from the mountain & that generally in warm weather for the sake of shade & water, nor has any of their young of yet ever been got hold of to be domesticated, although repeatedly attempted from the extraordinary care of the mother, living with her young in some retired peace till old enough to protect itself from being taken by surprise. By two the Welshman got some pieces of beef roasted on his sticks & placed it before us on a piece of bark, which with a little salt & our mouldy biscuits made us a comfortable meal, although the rest where it had been cut from was exposed in slices to dry in the sun on pieces of bark & the bushes within our sight nearly black covered with flies in appearance like carrion. While we had finished our repast, ending with dessert of strawberries gathered by the natives while our meat was roasting, Mr. Talbot & Wilson now retired under the shade of a tree to have a nap, having felt themselves somewhat tired. Having by this time placed all the specimens of plants collected since the commencement of our journey regularly in paper to be left till my return, I again went a short way in the wood to look for more for an hour, while my companions were enjoying their rest. On my return a short time after three o'clock I took the temperature of the air which was the same as at ten (69) coming through the wood. Our guide having recommended when we halted at noon, only to wait here but three hours as it would be necessary to travel six miles farther at least towards the mountain, to be able to gain the summit at an early hour tomorrow before the horizon rose to prevent us from seeing the ship at anchor in the harbour. The proposed time of our stay having already expired I disturbed all those of our party who were enjoying their nap to prepare for the next stage & shortly resumed our journey, but not without an accident which turned out afterwards with considerable disappointment, owing to the stupidity of a native dropping out of his hand a large calabash containing upwards of two gallons of strong brandy & water the remaining share of my spirits brought on the journey. We had travelled scarcely three miles when a thick fog commenced, over tufts of dry rough grass full of cattle tracks crossing in various ways. The soil was composed chiefly of sandy pulverized lava covered with numerous beds of strawberries & by the sides of several small ravines, made by the torrents of water from the melted snow running here at certain seasons, grew raspberries in great abundance & of a better flavour than those met with in the upper part of the wood, likely from no other cause than their being here more exposed to the sun. By six in the evening we travelled upwards of five miles when the fog became so thick as scarcely to be able to see ten yards before us, & ourselves drenching wet, shivering with cold & nearly beyond vegetation to afford us shelter for the night. We now began to consider it necessary to cut a few boughs from low trees of the *Acacia* & a species of *Sophora* growing on the side of the ravine to erect a hut, which in the course of little more than half an hour we accomplished, having a small axe & saw of mine with us, as also in getting plenty of firewood, which we kept burning within a yard to where we lay the most of the night. The lofty trees of *Acacia* noticed in the morning coming through the wood, was here stunted & decayed, scarcely attaining the height of ten feet, but in general the foliage was much broader. Although fatigued & wet I could not help, but ramble about among the numerous cranberry bushes picking specimens till prevented by darkness. In taking the temperature of the air at seven when almost dark I found that the thermometer had fallen to 52, which occasioned us all to feel cold & uncomfortable during the evening although our fire was good at the entrance of our hut.

17. While yet dark, ourselves cold & restless, by the recommendation of our guide, we got up at two am when the temperature of the air was 56 & at three commenced on our journey for the mountain leaving behind the natives who were all unwilling to accompany us for fear of cold. At five day-light began to appear having travelled about three miles over sandy pulverized lava, sinking every step over our ankles, full of cattle tracks & in places tufts of dry grass with a few low growing shrubs until within the last mile, which was destitute of vegetation, excepting one plant belonging to the tribe of *Syngenesia*, in habit of growth much resembling a *Yucca* with sharp pointed silver coloured leaves & a green upright spike to the height of three to four feet, producing pendulous branches with brown flowers, truly superb & almost worth the

journey of coming to see it on purpose. The majestic clouds rising on the horizon at day-light encircling us all round, as if it were an imminence wall with towers of various forms & sizes on its top placed at unequal distances gradually rising & keeping changing into fresh forms almost every second that had the finest effect imaginable. The temperature of the air had now fallen to freezing point (32) & for some time before Mr. Talbot & Wilson complained of being overcome by sleep & unable to proceed any farther feeling themselves quite exhausted. Our guide & myself with the two lads with us from the ship having waited by them for half an hour during which time we tried to rouse them both several times but without effect. Both myself & the guide being anxious to gain the summit so far as to see the ship when we promised His Lordship to make a fire by way of signal for him to see where we was with his spy-glass & finding it almost impossible to awake our other two companions we proceeded on our journey leaving one of the lads beside them & taking with us the other when after as we kept advancing every step now became steeper & more difficult to travel, vegetation had ceased. The yucca-looking plant no more appeared before us growing thinly scattered upon the lava. At six we succeeded in getting up the mountain to be able to see the ship which appeared no more in size to us than a vessel of about fifty tons, here we managed to gather a few of the decayed stumps & leaves of the yucca which we set fire to, remaining by it for half an hour anxiously looking after us for our two companions coming to overtake us, but seeing during this time nothing of them we kept on our journey at times over hard uneven brittle lava for some hundred yards & at others over sand which filled our shoes that annoyed us much. The mountain now became divided into several high conical sandy hills within places on their sides more than half way up to their top several old small volcano craters forced above the sand for some yards in height, & bleached nearly white from long exposure. The sun had begun to rise above the horizon & the air became more warm & pleasant, yet although obliged to rest to gain breath every other few hundred yards from the difficulty of travelling & the lightness of the air, it may appear somewhat remarkable that we felt no symptoms of perspiration, but only a difficulty of breathing soon after getting up to walk afresh when we had taken rest. At a little before eight o'clock while resting to breathe we saw a person following, which for some time we could not distinguish who of the three we had left behind it was till he approached nearer, then we found it was the lad left with Mr. Talbot & Wilson, whom we concluded might perhaps be sent by them to overtake us with some message, we therefore proposed to wait his arrival & in the mean time take a little refreshment, when at half past eight he came up with the greatest difficulty & then begged of me not to take him any farther, but allow him to return, he informed us that he had left both the others still asleep & feeling himself so cold sitting beside them he thought he might yet overtake us after having waited about an hour. When we had given the poor fellow a short time to rest while he [had] something to eat & a small drop of spirits to refresh him we sent him back to meet the others if possible giving him all the provisions, water & spirits we had excepting about a pint of water & a biscuit each & part of a bottle of port wine which was given to our guide by His Lordship to drink his health when we reached the snow. We also sent back [to] Mr. Talbot, his theodolite which he had brought for the purpose of taking the height of the mountain, if sufficiently early on the summit before the horizon rose to prevent his obtaining a base line, which already had been too late for more than an hour to gain such an object. The temperature of the air had by this time arisen to 46 & the sun shining bright, without any clouds near it excepting those that were beneath on the horizon, the air was dry & somewhat chilly. On resuming our journey by the bottom of the conical sandy hills, on our right & at times some distance up, upon their declivities the surface on which we travelled was every little while getting more oftener changed than before & much more uneven, having whiles lava, sand, small broken stones about the size of brick bates & at others have to scramble over large sharp edged granite stones of several tons weight resembling those that are blasted by miners, which have beyond a doubt been thrown up by some former convulsion. In this manner we kept slowly travelling on, rising very rapidly every step till we came in sight of the snow at a little past eleven, when for some time before our water was done & nearly all the wine finished by our guide who complained more than the lad & myself of being very ill with violent pain in his ears, which caused deafness & a headache, as well as an inclination to vomit almost constantly. We had now about two miles yet to reach the snow over a still more rugged & stoney surface, with a steeper ascent & ourselves fatigued & unwell from the same effects already mentioned, which on our getting within a quarter of a mile of the summit where the snow lay our guide at last was unable to proceed any farther, and prayed for god sake for us to make haste & bring him some snow to quench his thirst. The lad Mantle although young and slender held out better than we at first expected. At half past twelve I reached the snow which lay on the summit, chiefly on ancient porous lava of a sponge colour & in other places on sand, that was generally of a red colour intermixed with black & red cinders like the other conical hills we had passed. Some of these cinders had common quartz & two or three more kinds of minerals that I am yet unacquainted with very often bedded in one lump of lava. The snow in some parts was about three feet deep congealed into solid ice excepting from two to three inches on top of rough particles of loose snow. The whole appeared to be melting very fast, through the porous lava, as if through a sieve, which prevented us from being able to fill our empty vessels with its water, nor was more successful at where it lay on the sand. We therefore filled our handkerchiefs with the snow taking mouthfuls at the same time to quench our thirst, which we had some difficulty to keep within till it melted being so very cold. Having sent some to our guide by Mantle, while I remained to take the temperature of the air, both in the sun & the shade, which I found to be in the former 52 & in the latter in holes beneath the snow 44. I afterwards waited for nearly an hour travelling about to different parts admiring the scenery around me, which for twelve miles in every direction looked dreary & inhospitable, but the open pasture land below where the wild cattle feed had a pleasing effect. The wood still farther underneath adjoining it, which in a manner encircles the island of Owhyee was mostly hid in fog, which prevented me from seeing a greater distance than about twenty miles in a direct line underneath, but that which was of more elevation, such as Mouna Rou or the other smaller hills central in the island their black barren looking tops could be easily distinguished, nowhere covered with snow. During the time I remained on the summit I met with the entire skeleton of some animal, but without flesh and apparently not long dead, which on returning to our guide was informed that frequently the numerous wild dogs in this part had already nearly destroyed what sheep had increased from the few brought by Vancouver at the same time when he brought the cattle being often pursued this far beyond vegetation where [they] got bewildered & soon after died for want of food. Being now after two o'clock & feeling ourselves still unwell with the same complaints as before & unable to travel but slowly homewards we left this interesting place falling rapidly often scrambling with difficulty over the sharp edged granite stones, we had before so much trouble in ascending over. Our small collection of minerals & a few other things brought with us felt nearly double their real weight as we kept advancing home, & the loose sand which got constantly in our shoes, wore our stockings full of holes, which made our feet so tender that when we had to travel upon anything hard, it was with the greatest pain we could bear our weight. In this hobbling manner scarcely able to drag our limbs for the last four miles, we reached our hut at half past six, much recovered from our former complaint, where to our surprise we found on our arrival that the lad sent back in the morning did not meet with Mr. Talbot & Wilson. At this unlucky circumstance we both felt not a little uneasy knowing them to have neither food nor water to refresh them during the day & night now fast approaching & them unacquainted with the way to find us out, & we even farther doubted at times whither they had not felt bad effects from their falling asleep at the time we left them, which we now became sorry we had done, fortunately however our fears was soon relieved before half an hour, on hearing them calling out aloud not far distant. When they arrived they did not appear so much fatigued as ourselves notwithstanding their want of food & water all day. When we enquired of them how long they had slept, they would only acknowledge somewhat better than an hour & that when they awoke they followed us but not finding any traces of our footsteps they soon gave up the hopes of overtaking us & made for the first of the highest hills which we had passed in expectation, when they gained the summit, to see the snow, but in this they were sadly disappointed after much fatigue from sinking they said at times up to the knees in sand. The snow which we had brought home with us served fortunately to supply the want of water, the natives having in our absence drank all we had left to about a pint, which if it had been my calabash of spirits & water [broke] yesterday they would not have attempted to make so free with it, as they had done with our water. During the evening (which was to-night free from wet fog with only a light dew) we all enjoyed ourselves very pleasantly staying by the fire in front of our hut, melting the snow in my specimen tin box placed far enough from the fire not to get too warm, the duty of which was performed by Mr. Talbot who gave us at our supper, each in his turn, a small quantity at a time, regretting himself much that he had fallen asleep which prevented him from seeing it on the mountain or even meeting with us on our return shortly after we had left to put him in the way to see it. The natives as soon as we had seated ourselves after coming home, began to rub our thighs & legs with their hands, a custom very common practiced among themselves when they feel tired after walking or at any other time when they feel pain in their legs or arms, & term it, Lummi Lummi, which gives the parts they rub a warmth soon after & causes the stiffness of the legs to feel much less while they are rubbing, which is pleasant & very agreeable. The temperature of the air at seven pm was 50 at 10 48. The summit of Mouna Kaah from Byron's Bay (or Heddo) we calculated to be about 70 miles by the common path that is frequented to come above the wood to kill wild cattle, but in a direct line perhaps only one half the distance. At Lapahoi 40 miles to the westward of Byron's Bay where we commenced our journey nearly in a direct line under the mountain rising grandly with a gentle ascent all the way till within the last 12 miles of the peak, the height of which we judged could not be under 18000 feet above the level of the sea. The land along the sea coast from Byron's Bay to upwards of 40 miles to the westward & to about six in breadth was free from wood excepting by the sides & bottom of ravines. The wood that surrounds chiefly the central part of the island commences here at the distance of five or six miles from the sea coast which continue to the breadth of 12 miles, intersected with numerous deep ravines & large rivers of the finest water. The outskirts next the sea coast are chiefly *Metrosideroses* & *Aleurites*, of the former there are several handsome coloured flowering species, which entirely disappears after five or six miles towards the centre of the wood. The one which is the most commonest of this tribe attain often the height of 40 feet & is thick in proportion much used by the natives for buidling their huts, being hard & durable. The upper part above the woods resemble pasture land for seven miles farther & is thickly covered with low growing shrubs & abundance of strawberries & raspberries. Afterwards vegetation soon ceases for the last eight miles towards the summit. The clouds generally rise on the mountains of Owhyee & the other islands in the morning & again disperse towards the evening, when commonly more or less rain falls during the night & often in the day at some distance off from these peaks, when on the coast the sun shines without any appearance of rain.

18. A fine morning. Got up at day-light after resting but indifferently from the frequent howling of wild dogs in the night & excessive cold which occasioned us to get up often to keep the fire burning. At six o'clock we set out on our journey homewards for the two foreigners' hut by the edge of the wood, nearly six miles distant, but had scarcely left our hut many yards when the natives unknown to us set fire to the hut which soon destroyed it. A common custom our guide informed us strictly adhered to by these people when they happen to flit from one place to another. At 10 we reached the Prussian and Welshman's hut, without seeing since we had left them any wild cattle but only their footmarks very frequently & one wild dog & a cat. The dog seemed to be of the same kind as the native one domesticated which they eat the flesh of. The cat appeared to be like the European. Shortly after our arrival we had plenty of slices of roasted beef for breakfast & abundance of water which took us some time before we could drink enough to satisfy our thirst. After

breakfast my companions spread their mats under the shade of a tree, where they slept till noon, an indulgence which I could not afford, from my having to shift the specimens collected since I landed after the commencement of our journey into dry papers. By one pm I got all ready & others gathered as well as strawberry & raspberry plants for to take with me to England. At two the whole camp was all upon the move to the sea side at Lapahoi where we ordered the canoe & natives to wait for our return when landed & by whom our wide awake purser (Wilson a true N.B.) now intended to forward all the fresh beef both these men had for the use of the ship's company. It was astonishing to see the heavy loads the natives with these men willingly undertook to carry a distance of 25 miles at least. Some of them had a whole side of a bullock & scarcely anyhow less than a half. We on our departure formed an amusing medley of beings on our going through the wood in a straggling manner at some distance a part after each other in a kind of narrow path when we had but a short way travelled when the afternoon became very wet, so much so, that for the last four miles we could hardly go a step without falling on the roots of trees & mud, much to our annoyance. When we had got about half way through the wood my three companions & the two foreigners left myself & the two lads behind with two of the natives, who on account of our loads of plants &c. could not keep up with them, & did not arrive till dusk at the first hut, where we all slept at the first night going up to the mountain, which on our arrival drenching wet we found only the two foreigners, as the others had kept on for Lapahoi. Both these lazy fellows to whom I gave a dollar each when they had left me in the wood promising to have a good fire to dry our clothes & something warm ready for us to eat by the time we arrived, had not yet thought of anything of the kind although at home for some time. However it was the best policy not to grumble at their neglect, but keep on the best terms possible. By nine o'clock we at last had some stewed beef from them, which was cooked in an old saucepan & afterwards placed between themselves two & myself, giving the lid with some in it, to the two lads, which when we had done went soon after to sleep with our clothes scarcely half dry on a dirty mat in a corner of the hut amongst a houseful of natives & stinking beef. Thermometer at 5am 43 at 10 63, at 12 75 & at 7pm 66.

19. A hazy morning with some light showers of rain. Got up at day-light & felt so tired that I could hardly walk. Being Sunday the blacksmith informed me that the natives with me were afraid to accompany me with their loads

to join the other party before, whom I was now anxious to overtake. Having enquired for what reason they declined travelling on Sunday I was told it was a religious form put in force by the missionaries, like the others prohibiting them from lighting fires on this day to cook their meals & of bathing themselves. At eight had some breakfast, which after I had done I now urged the necessity of joining my companions before noon, & requested of the blacksmith to get the natives on the way as soon as possible, which he promised & would get his own people with their loads of beef as well & then himself would accompany me at eleven o'clock when likely by that time the morning would get free from rain. Having waited to the hour he promised without any signs of his coming with me from a party of the natives with whom he had began at an early hour in the morning to gamble with for tapa cloths against part of his stinking beef who the natives eat with equal fondness to its being fresh. The game (Nooah) which they played at, is one of the most ancient kind with them & one which is most frequently seen played at among them. This game consists of five small tapa bags stuffed with cotton or the down of ferns, placed in a row beside each other, underneath one of which are placed a stone, so as to deceive the parties playing, which of the bags it was put under, the players are seated round & form a circle with each a small wand in his hand with which he strikes the bag that he supposes the stone to be under. There are generally ten in number playing at the same time with different coloured rags tied to the upper part of their wands by [way] of distinction between the parties. I have been creditably informed that at this game they gamble their hogs & every other thing they possess, nay even their wives & are always strict in paying their debt of honour to the winner with equal punctuality of a Newmarket sportsman. Shortly after eleven while it still rained lightly, I left these people gambling, for Lapahoi with the two lads & two of the natives, when at a little before three in the afternoon I joined my companions, who I found had expected me in the morning & were now about to leave for home by land without me, the canoe having returned the second day after we had left. At seven in the evening after crossing a number of deep ravines & wading through rivers at times to our middle with only our shoes off to make a sure footing over the slippery stones in the bed of the river, we took up our night quarters at a dirty hut, drenching wet from the rainy afternoon & the numerous rivers we had crossed, having travelled about 10 miles east of Lapahoi. While shifting ourselves with what little spare dry clothes we had, the natives lighted a fire & went about cooking a few cabbage, bananas, & bread fruit given us on the road, the latter fruit I recommended to be roasted as Mr. Talbot & Wilson had never yet tasted any. The manner in which the natives cooked in their own way the others afforded us an opportunity of seeing something new & curious worthy to be related. Their first process to get some stones heated red hot, then after put into a gourd calabash with a few leaves of the tea tree underground, & the same above again, on which the cabbages was placed cut in pieces intermixed with the bananas, who had only the outer skin taken off. Then over all at top more leaves were put to prevent the steam rising upwards. In about a quarter of an hour all was ready, when the calabash was placed before us, & we then commenced eating, using our knives as a substitute for forks & found our vegetable supper by no means unpalatable cooked in this simple manner. Our own stock of provisions being now reduced to only three biscuits & about a pint of brandy left in Mr. Talbot's flask, which we found very acceptable when taking a mouthful to warm us, while we had wet clothes on. About nine o'clock we spread our mats in a corner of the hut & retired to rest, promising each other, whoever woke first in the morning to rouse the others that we might leave early to continue on our journey let the weather by wet or dry, as nothing could be got from the natives to eat for anything we have to offer them, being in number eleven & yet a long day's journey from the ship with numerous deep ravines & rivers still before us along the coast. Thermometer at 6am 66 at 12 73 & at 7pm 72.

20. A cloudy morning & rather cold. Got up at 4am & shortly after commenced our journey crossing every other short distance deep ravines & rivers without waiting to take any of our clothes off, but merely our shoes. One of these large rivers we had to cross at seven o'clock nearly proved of serious consequence to Mr. Talbot, while attempting to cross in the same part with the natives against their will, as he was unacquainted with the stones where he ought to put his feet upon as they were themselves. Our guide who could not persuade him from it like the natives, went with Wilson higher up to cross in another path while I remained fortunately with him to see him cross assisted by the natives, when on his going a few steps in the water he was instantly swept off his feet, but luckily he at the time got hold of a large stone that was above water with two of the natives upon it, who caught hold of his hands & a third in attempting to lay hold of [his] body, was before he could accomplish it carried away by the current to some distance not far from a high waterfall which I trembled with terror that the natives might perhaps quit their hold, & allow him to get with the stream. By this time he could only keep his head above water & out of my power to render him any assis-

tance without being in the same danger with himself & was impossible for the two who had hold of him to drag him out, on account of one of his legs having got jammed between two stones, which as they kept trying to put him out of the water his leg he was afraid would get broke. At this misfortune & seeing myself of no use to him I waved my hand to the others crossing above to hurry them on the opposite side, while I managed to get near enough to get the barrel of my gun underneath his leg & fortunately soon succeeded in extricating it, when the natives after had him instantly out, just as our guide & Wilson arrived without farther hurt than being bruised a little & his watch partly injured from the water having got inside. The frightened natives had shown so much anxiety for his preservation that he gave each of them on being out of danger a knife as a mark of gratitude for what they had done. These poor fellows during the remainder of the day often talked of his narrow escape from being drowned, which if it had happened they said Kaumanna would be sure to kill them for not saving his life, so that they would sooner have died with him than go home alive. By the time the accident was over Mantle & the lad with the rest of the natives came up, who after some trouble got all across safe, when we then again continued on our journey & soon after came to another river nearly the size of the last, which in crossing Mr. Wilson fell & broke our flask that contained our little brandy left. At nine o'clock we halted at a large hut pleasantly situated among a grove of bread fruit trees, where we expected to procure something to eat, but in this as at other times we was disappointed & only get a few tarro & some poi, which we was glad to make a breakfast of & afterwards continued our journey crossing ravines four or five hundred feet deep & drinking the fine clear water of almost every one we crossed. These ravines had on their sides in most places plenty of the rose apple & bread fruit trees in fruit, with numerous ferns & other plants, besides *Pandanuses* forming woods of themselves in many parts as we passed. At five in the afternoon we arrived at last opposite the ship much fatigued from having to get up & down 70 ravines since we had left Lapahoi on Sunday & travelled a distance of more than forty miles. Mantle who succeeded to go with me to the snow & two of the natives was the only three who managed to keep up with us, the other lad with the rest of the natives were left behind many miles. When we had waited for some time to procure a canoe to take us over the last river at the watering place, then afterwards to put us on board, where I was asked by Mr. Talbot & Wilson to dine, having scarcely tasted food since I left the blacksmith on Sunday morning & who is now with the Welshman on their way to Byron's Bay with their ill-flavoured beef. On our getting on board I learned that my things on shore remained yet where I had left them & my friend Mr. Forder still living in the old tent. I therefore got away unperceived by Mr. Talbot and Wilson without having anything to eat or drink & got with my man Mantle into the same canoe that took us on board to put us again on shore near the tent where I found Mr. Forder who informed me that he had repeatedly applied for a boat to remove my things to the chief's hut on the opposite side of the bay where permission was given me to stay at. When I had got myself shifted & had something to eat such as my salt provision from the ship could afford, His Lordship sent for me to drink tea with him, which I did not from being too much fatigued & wanted rest, yet the surgeon & A. Bloxam who are both almost constantly with him, came afterwards in the evening & tormented till after nine o'clock to hear the accounts of our journey. Thermometer at 5am 66 at 12 78 & at 7pm 74.

21. A fine day. Shifted my specimens during the morning into dry paper. At nine o'clock the lad Trounce with the other natives left behind yesterday, arrived with the rest of my specimens & seeds &c. who on getting off his load begged I would not be angry with him for telling me that the natives upset the canoe with himself & all my things, when coming over the river at the watering place & had nearly drowned him, on hearing this accident I felt rather mortified for fear that the salt water might have destroyed my seeds, but Mr. Young being present with me at the time who assured me that the water could not be scarcely brackish at the time he had crossed as it was near then low water. With this consolation I immediately went about shifting everything into dry paper which while thus busily employed Mr. A. Bloxom came to get a sight of the minerals I had collected which I readily granted, but could not get free from him till I had given him some. A short time after Lord Byron came & said I wanted to shun him last night & this morning for not coming to tea & breakfast with him. He farther said that as the surveyor had come up from Woahoo he would want the tent, to which I made answer that, I intended to go as soon as I had my plants & seeds put in dry paper to hire a hut while the ship remained. In the afternoon got Mr. Young to accompany me as an interpreter to hire a hut which we soon succeeded for two blankets for three weeks in preference to money & would be ready for me tomorrow forenoon, so that my friend & myself will once more have a hut of our own as we had on the island of Woahoo. After dusk heavy rain which continued all night coming through the old tent where we lay on the ground. Thermometer at 5am 74 at 12 84 at 7pm 82.

22. A fine morning. Went soon after day-light to see if the people whom I had hired the hut from had begun to remove their things to be ready for me at the time they promised. In the forenoon upwards of 200 of the natives came from the eastward part of the islands with their rents for Queen Kaumanna, bringing her such as tapa cloths & mats of various dyed patterns & hogs, sweet potatoes, tarro, bananas, sugar canes, coco-nuts &c. When they had got in presence of their Royal Mistress, who I noticed never attempted to smile or give a pleasant look to any of them, when she examined what each of them have brought, although I was told she had not been on the island for the last four years. Some of the sugar canes were large & well grown, but did not possess any variety from what I had before seen in the West Indies, three of the largest. I asked permission to have which was readily granted. The surgeon, both the Bloxams & draftsman were looking out for something of another kind. A mat which was given to the former gentleman while I was present with old Young I was informed it would take one woman twelve months to make it. It is the duty of the women when they get to a certain period in years to manufacture all the tapa cloths & mats as well as many other kinds of labour. Being busily employed in removing my things to my new hut at the time, I was obliged to leave so interesting a group of people come into their Queen with what little they possessed to satisfy her for their rents. When I had got with the first of my things to the hut & on my offering beforehand the two blankets for payment of rent, as I was told by Young was customary with themselves, the owner of the hut would not accept them from me & seemed unwilling for me to come & his wife with some others of their family present shed tears which rather surprised me to know what occasioned them to alter their minds so soon. On my returning for the rest of my things I again brought with me Mr. Young to ascertain what I had done to displease them, & found it was no other complaint than their fear of my burning their hut when I went away, as was customary with themselves. The owner of the hut & his wife being assured no such intentions was ever thought of, they now at once received me as their tenant without fear of doing their hut any harm. In the afternoon shifted my specimens on my chest, having no table or anything else to set upon. At dusk the chief man of the place came at Mr. Young's request to taboo my hut to prevent my being crowded by the natives. The simple form of this charm gone through by the chief, was his sticking in the ground a few wands at some distance from each other a little way all round the hut, which on being seen by the natives they never durst to offer to venture inside to this door of the hut unless they are sent with some message

by any of the chiefs to the person whom it belongs. Thermometer at 6am 74 at 12 85 & at 7pm 82.

23. A fine day. Went soon after day-light about nine miles to the eastward of Byron's Bay to collect plants. At three in the afternoon returned, having met with little success, the plants being mostly *Metrosideroses* & *Pandanuses* with some ferns all growing upon broken lumps of lava. On my return home I was informed that Mr. Talbot & Wilson my former companions to Mouna Kaah had called to acquaint me that it was their intentions to go to the volcano on the 25th & wished to see me in the evening on board to arrange matters with them as it respected a guide & people to carry our provisions &c. At five went on board & got our arrangements settled, then returned on shore at dusk when shortly after it commenced heavy raining till after 10 o'clock. Thermometer at 6am 74 at 12 82 & at 7pm 78.

24. A fine day. In the morning shifted my specimens, & then after breakfast went to engage the blacksmith met with at Mouna Kaah as guide, also five natives, who he recommended me to speak to Kaumanna to get her to order them not to leave us till our return as otherwise they would likely disappoint us on the road. During the afternoon got thing[s] necessary prepared for our journey in the morning. Thermometer at 6am 76 at 12 84 & at 7pm 82.

25. A fine morning with some light showers of rain. At 6am met Mr. Talbot and Wilson & the guide, but none of the natives hired to go with us had yet appeared, our guide was therefore obliged to go after them to their huts to get them on the way, which they are always very reluctantly to do early in the morning from the effects of cold. While we was thus waiting for the natives we had the opportunity of seeing both the Queens & one of the gentlemen living on shore with His Lordship bathing together, they with their short petticoat on & him the maro. By seven our guide had brought the natives, when we soon after commenced our journey as the first party for the volcano from the ship, His Lordship & several others intend to follow us in a few days. After we had travelled about three miles we halted to have some breakfast, then after resumed our journey for ten miles farther when we again stopped at one pm to have some more refreshment. Part of the road which we had travelled was far upwards of five miles through a wood over a narrow path of broken pieces of sharp edged lava, which we could hardly bear our own weight upon without pain to our feet. This wood although growing entirely without

any signs of anything nutritious to support even a shrub, yet many of the trees particularly the *Metrosideroses* attain the height of more than 40 feet, but are slender in proportion & far inferior to their size in the woods going to Mouna Kaah, under their shade grow everywhere a number of ferns which hide the lava from the eye & touch each other on both sides the path nearly all the way through the wood. At six in the evening we reached the first hut newly put up for the accommodation of Lord Byron on his way to the volcano, having travelled about 20 miles since we had left the ship. The last six miles of our journey was through an open country over solid greyish black undulated lava, covered with numerous short stumpy ferns chiefly of the *Cyathea* tribe, which the natives often burn during the dry season, we met also quantities of three sorts of cranberries on each side [of] the path perfectly ripe, which we eat freely [of] & enjoyed them much although they were somewhat acid to the taste. The wood which we had come through in the forenoon continued at some distance on our right all the way up in the country toward the mountain of Mouna Roa. Having now taken up our quarters in the new hut for the night, we were soon joined by a great number of the natives come from the neighbouring huts, who many of them eagerly approached us to examine our hands & feet without the least diffidence, to ascertain if the skin was of the same colour with our face, but the children who were under ten years of age a look from us would make them take to their heels directly trembling with fear. We noticed like many other times that the old women generally nursed the children who they carry on their backs whither they are either walking or sitting & is never seen to have them in their arms. When they have occasion to travel some distance the women then tie their children round the neck & thighs on their back with a bandage of tapa cloth, which gives the woman the liberty of her hands to carry something else. At dusk a thick fog came on with light showers of rain. Thermometer at 5am 73 at 12 82 & at 7pm 66.

26. A wet foggy morning which prevented us from resuming our journey till after eight o'clock. By noon we reached the last native huts on our way to the volcano, having travelled about 10 miles over the same kind of undulated lava as on the afternoon of yesterday. At this place we remained till our guide went to the different huts to procure some fowls or a young hog to take with us, but after his trying in vain for two hours to bargain with them he could not succeed although they had plenty of both which they would not part with unless we gave them three times their value, we therefore continued on our journey with what little provisions we had of our own, but not without much

grumbling on part of the natives with us, who felt themselves disappointed for not getting a fresh supply of tarro & would they said have to cut ferns before we returned. However after we had parted from the others at the huts some distance they sent after us a native with a couple of fowls for which they refused to take any payment. At six in the evening we reached two old small hovels at the upper part of the wood chiefly trees of the *Acacia* which they make their canoes of. Having travelled about 18 miles since the morning over lava covered with ferns, cranberries, & low straggling bushes of the pretty red flowering *Metrosideroses* with a variety of other plants & some dry tufts of long grass, besides numerous beds of strawberries when we approached the last two miles growing underneath the ferns & grass upon sandy pulverized lava ashes of no great depth from the same sort of undulated lava which we had been travelling on before. By the recommendation of our guide we here took up our quarters for the night, being given to understand that we was now not far from the volcano, & that we could not get so good a place farther on to remain at for the night. We therefore immediately set to work to repair the huts before dark with fern leaves, while some of the natives made a fire to cook the fowls given us on the road which with some salt beef made us a comfortable supper, having had nothing during the day to eat since morning. At the time when we were getting fern leaves in the wood to repair the huts, our guide left us for a short time without our knowledge & on his return informed us that he had seen the volcano smoking at no great distance, but having doubted his report being truth from his manner on the journey which often gave us reasons to place but little confidence in one half he said, it by this time getting dark none of us felt inclined to leave the huts, which we sadly repented the following morning. Thermometer at 5am 58 at 12 76 & at 7pm 60.

27. A misty foggy morning which prevented us from getting on our journey till seven o'clock when after we had taken our breakfast & sent one of the natives back to the first huts to have what he could get for us prepared by the time we returned to him at night. Soon after we had commenced our journey we passed on our left a large old volcano crater now closed & covered with verdure on its internal declivities & over the bottom which is level & more than 1000 feet deep, within places a few low growing red flowering *Metrosideros* bushes. At eight we reached a low shed on the brink of the present volcano in action which we with ease could have got to last night, without having afterwards any trouble to repair it, being in good condition & besides we

Great Volcano of Peli, at Hawaii. From Maria Callcott's *Voyage of H. M. S.* Blonde *to the Sandwich Islands, in the years 1824-1825. Captain the Right Hon. Lord Byron, commander.* "Engraved from a Drawing made on the spot by Robert Dampier."

would have the gratification of watching the motions of the burning craters underneath during the night which we now lamented to have missed from the ignorance of our guide not knowing the place, as we on the journey suspected. At this hut we stood for half an hour gazing with astonishment on the immense depth below covered with clouds & smoke, while at short intervals a terrific noise was distinctly heard among the different numerous burning craters. The natives while we remained were busy making sandals of grass to protect the bottom of their feet from the lava when they got below, where they felt inclined to accompany us without being invited. At half past eight we all began with eagerness to descend from the hut with walking sticks in our hands, having nothing to hold by but short tufts of dry grass bedded in sand & loose stones which kept rolling amongst us all the way to the first ledge of lava where we got upon at half past nine o'clock without receiving any injury further than a few bruises from the stones. The vegetation here ceased & the burning craters still at least 500 feet below us & our road more difficult & steep over large irregular sharp edged stones from their bed piled up loose upon each other in various shapes, intermixed with flat pieces of honey-combed lava. By noon we had some of us succeeded with great difficulty & danger to reach the nearest smoking pillar about 30 feet high covered with sulfur which gave it a beautiful yellow appearance, here we remained for some time for those of the natives behind had come up, some of which were lame & had their naked legs bleeding from falling on their way through the hollow lava that lay in places resembling flues on the top of the other more solid underneath, which required the greatest caution to try it first with our sticks & if it broke before we attempted to advance [to] step upon it. One poor fellow who was carrying a calabash of water which he broke, lamed himself so much that two of the rest was obliged to assist him off the lava & return back, having got besides his lameness much frightened at seeing the numerous wide vents before him that we had crossed. Some of these openings were constantly smoking & smelt so strong of brimstone which got up our nostrils, when going over them nearly to suffocate us. While we remained by the brimstone pillar, we noticed it issued out at intervals sudden strong gusts of smoke strongly impregnated with brimstone, which obliged us to be careful to avoid when near it picking up specimens of the hot lava covered with sulphur from the openings in the sides. The natives who all of them experienced both fear & difficulty in coming this far, wished now to return again to the top which was readily granted when ourselves four went farther on to examine another pillar about 50 feet high burning red at the top like a furnace

& emitting to a considerable distance from it lumps of soft lava & at times cinders of various colours, which on our getting near it the terrific noise kept up underneath everywhere round it and a considerable way from where we stood intimidated our fears while we remained that the part under us might perhaps be torn to pieces every minute, such as the frightful loud noise the burning underneath kept up, which lowered at intervals for a few seconds among the numerous pillars, who were nearly hid in smoke & impossible for persons to approach them from the strong smell of sulphur, excepting to windward. The burning noise of the craters resembled a little the forge of a blacksmith & could be distinctly heard half a mile from them. At one pm we began to return to ascend to the top a mile farther to the northward of the hut where we descended from, as our time could not afford to remain any longer at this interesting hot place, where we found the temperature of the air to be in the shade 92 & in the sun exceeding the height of my thermometer which was only 132. The volcano is situated about 40 miles south east of Byron's Bay, at the foot of Mouna Roa next in height to Mouna Kaah. Its circumference, according to the statements of the missionaries, who had measured it some years ago, is seven miles round the edge at top, & below round the basin five, which is almost circular. The depth in most places is from 1000 to 1200 feet, the ledge which is nearly perfect all round the basin, is rather better than half way down from the top. The number of burning craters in form of pillars exceeded twelve, some of which were upwards of 50 feet in height & generally of a bronze shining colour of various shapes, with several unequal openings on their sides to the summit, where sudden gusts of smoke issue out from them at intervals. The surface adjoining the craters varied in many places where we had travelled. At times we met with sharp edged granite stones piled up above each other often intermixed with other similar hard kind of stones of a red colour & some smooth & white & others dark, but in general the basin was composed of black & greyish porous honeycombed brittle lava, often undulated like the waves of the sea & where late convulsions had taken place it formed at every other short distance hollow burning passages for a great length, which are full of red & white ashes & when the thin crust above the passages are trod upon, it easily falls in & intimidates persons walking upon it from the constant sight of having to step over the numerous deep smoky vents at no great distance apart. The most of the lava was crystalized in small particles on the top like glass, which if fallen upon tore the skin of our hands & feet through our clothes & instantly brought blood. The internal declivities on the west side is steep & free from vegetation, but on the south & east side vegetation continues in most parts to the first ledge, which this far is chiefly the common red flowering *Metrosideroses* now got quite dwarf & tufts of two or three kinds of dry grasses. To the south near the top we noticed a large space covered with brimstone, which reminded us of the chalky cliffs at Dover & Gravesend. At 2 o'clock we gained the summit on the north side where I met with several bushes of Sanders wood in flower near the top with a few other plants, but not the same variety as met with when near beyond the vegetation going to Mouna Raah, on the top were several hot springs, underneath narrow deep openings divided in the earth at short distances apart from each other for half a mile in breadth from the edge of the volcano at top. These openings had on the surface at their sides for more than a foot wide luxuriant patches of green moss growing from the moisture occasioned by the steam, which kept constantly smoking & when we held the palms of our hands over it, large drops collected upon them in a few seconds. Some of the water which we drank lying near the surface was only warm. Here we saw also two low temporary sheds, where the natives come to from the woods when they are cutting large trees to make canoes, for the convenience of water & cooking by steam the trunk parts of certain ferns (*Cyathea*) which they very often eat when the food brought them is done. Several pieces of this fern we found at the huts ready cooked, which the natives were glad to find & eat of fully & was by no means unpalatable. By 4 in the afternoon we again came to the huts where we stopt at last night, about a mile distant from the volcano, having met with on our return great quantities of cranberries & strawberries in fruit, which for want of time we had no opportunity of gathering, but very few. Soon after we had arrived on opening our calabashes to have some refreshment we found that the native left behind in the morning with orders to go back to the first hut to have something provided for us by the time we came up, had made free with the greatest part of our biscuit, which we could carry now in our pockets & made use of the calabashes for our tender brittle lava specimens. Some of the sulphur which I had put in a seed paper when down in the volcano at the first pillar, had already burned through both the paper & my coat pocket. Its taste was more of acid than of sulphur. When we had finished packing up our lava specimens as secure as circumstance would allow, we then after commenced our journey homewards for the first huts, where we arrived a short time after dark, & found the native waiting for us with some poi & tarro, being all he could procure for us, having mixed the poi with a little sugar, which Mr. Talbot and Wilson had brought with them for that purpose to take away its sour taste & warming the

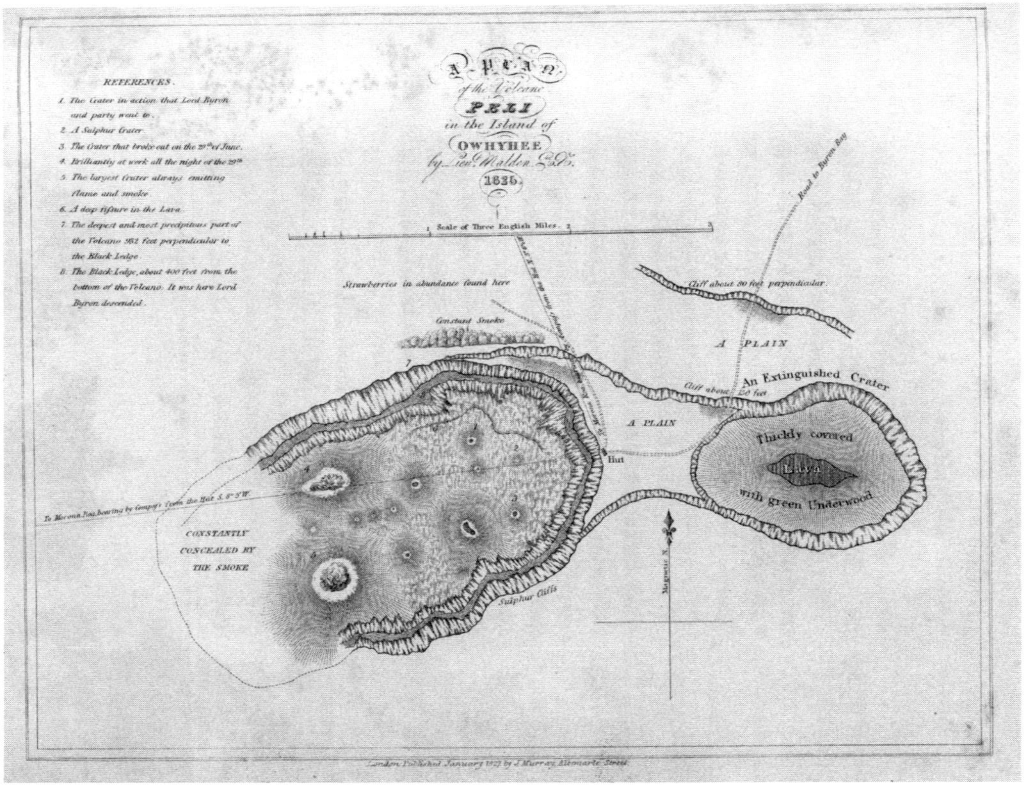

A Plan of the Volcano Peli in the Island of Owhyhee by Lieut. Malden, R.N. 1825. From Maria Callcott's *Voyage of H. M. S.* Blonde *to the Sandwich Islands, in the years 1824-1825. Captain the Right Hon. Lord Byron, commander.*

tarro before the fire, we managed to satisfy our hunger & soon after went to sleep somewhat tired, with intentions of rising at day-light to get home tomorrow evening. Thermometer at 5am 58 at 12 92 & at 7pm 64.

28. A fine morning, but somewhat foggy. Got up at day-light & soon after resumed our journey, which on travelling about four miles we was surprised to meet so early the advance party of Lord Byron going to the volcano, consisting of about 30 natives with the head chief (Maro) at Byron's Bay as their leader to have fires lighted to cook for His Lordship what was necessary by the time he arrived at the intended places to stop at. [Soon] after these we met the naturalist & one of the missionaries & a little way after them the surgeon & draftsman, then the chaplain with two stout natives, carrying his hammock slung to a pole ready for him to get into when he felt tired. Again after these were coming a short distance the surveyor, first Lieutenant another of the missionaries & a boy midshipman, with His Lordship still a short way farther in the rear accompanied by another boy midshipman & who had his hammock with him ready for to be carried like the chaplain. When we informed [him] of our want of provisions, which occasioned our return so soon, he requested that we should take what fowls or any thing else we wanted from the natives with him as we passed, who had plenty of every thing carrying for his use. Although His Lordship had been very lame for the last fortnight from the kick of a horse on his leaving the island of Woahoo he kept up on the journey with better spirits than most of them. The first Lieutenant had at the time we met him got already his feet through his shoes with a handkerchief tied on to keep his toes from the lava till he got another pair sent, after we got home. The cavalcade of natives attending on His Lordship, which we was particular in reckoning their number exceeded 200 composed of men, women & children loaded with all kinds of provisions, straggling one after the other in a narrow path that would only admit of one person, who occupied a line nearly a mile in length, which afforded us when passing them amusing sight to see so many half naked beings, some with a dozen of fowls tied by the feet hung on both ends of a stick thrown across the shoulder like the rabbit men in London, while others had a hog tied by the four feet thrown over the neck carried on the back & some had them tied with a rope drove before, when more than a hundred others were loaded with calabashes of various shapes & sizes containing poi, water &c. & a number of others carrying bundles of fish & tarro, sweet potatoes &c. tied up in the leaves of the banana & the tea tree. At eight o'clock we reached the hut where we slept

at the first night on our way to the volcano & remained till after nine, to roast four fowls given us by His Lordship & a few sweet potatoes. The natives with us on this journey like the others when going to Mouna Kaah we noticed eat the entrails, head & feet of the fowls after warming them a little before the fire, while the rest was getting done for us. The manner in which they cooked our fowls, when they had picked off the feathers, was afterwards dividing them into four quarters, each of which they put on the end of a stick placed in the ground in a leaning posture over the fire. About eleven in the forenoon we met a party of the midshipmen in the wood about seven miles on their journey to the volcano, with the Welshman met with at Mouna Kaah as their guide. At three in the afternoon we arrived opposite the ship much fatigued from our foolishly walking so fast during the heat of the day to be home at this hour on account of a trifling wager made with some of Lord Byron's party in the morning, having travelled 40 miles in 12 hours over a narrow uneven lava path which destroyed all our shoes, that had been new when we left. At dusk the natives who we left behind came home with our luggage very tired & lame. Thermometer at 5am 62 at 12 82 & at 7pm 75.

29. A fine day. During the forenoon laid my specimens of plants brought home in paper, & in the afternoon sorted & packed up my specimens of lava, as also arranged & shifted others gathered before I left. Thermometer at 6am 76 at 72 87 & at 7pm 80.

30. A fine day, but exceedingly warm. Employed all day among my specimens & seeds. In the evening died at a neighbouring hut a man about 30 years of age, which on my going to see him, found that the relatives had taken him out of his hut to breathe his last in a temporary shed opposite covered with coco-nut leaves, owing to a superstitious custom among them of never afterwards inhabiting the hut which any one by chance might happen to die in, but immediately burn to destroy it. The body was lying on a dirty mat rolled up in a piece of tapa cloth with a few of the natives over it making a howling noise & when I stood near them enquiring what was the matter they leaned their head on one hand, while they pointed the other towards the sky, where they said he had gone to sleep & parted from them forever. Mr. Young who was in the habit of calling upon me frequently in the evenings, came to my hut soon after I got home, & to whom I communicated what I had seen & when at the island of Woahoo of two others with the natives round them in a similar way, but had never seen yet a funeral since on the islands, to which

he informed me that he never had but one opportunity of seeing one himself, & that through a great favour, which took place in the dead of the night by a few old men & women who always are entrusted to bury the dead about this hour with great secrecy for what reasons he never could learn, only supposed it to have originated from an old superstitious custom. Some of them he said never buried the dead, but concealed them for a certain time till the flesh decayed off the bones sufficiently dry to be tied up afterwards inside the roof of their huts wrapped up in a bundle of tapa cloth. Thermometer 6am 74 at 12 90 & at 7pm 83.

1. (July) A fine day, but exceedingly warm. Employed all the day like yesterday among my specimens & seeds, writing &c. During the afternoon His Lordship with the rest of his party returned from the volcano. Some of them got so lame that they did not arrive till dusk. Thermometer at 6am 76 at 12 90 & at 7pm 79.

2. A showery morning. Went at 5 o'clock with Mr. Talbot & Mr. Goodrich our Mouna Raah guide to see the waterfall above Byron's Bay about six miles in the wood & upwards of 60 feet in height. The river is amongst the largest on this side of the island, which like the rest has many waterfalls upon it at no great distance from each other. At 2pm my companions left me to be home in time to dinner, while I with Mantle remained in the wood till nearly dusk gathering plants, some of which I had not seen before, particularly a beautiful climbing plant resembling a *Erythrina* with crimson flowers and several *Metrosideroses* and ferns. On my return home after dark I was given to understand that His Lordship intends to sail on Thursday the 7th for certain, which prevents me going again to the woods, from having a number of specimens & seeds to dry & arrange that cannot be accomplished on board for want of not having a place of my own. Thermometer at 5am 74 at 12 86 & 7pm 80.

3. (Sunday) A fine morning, employed all day among my specimens & seeds, writing &c. In the morning another party of gentlemen from the ship gone to the volcano to be back in time before the ship sails. Thermometer at 5am 74 at 12 86 & at 7pm 79.

4. A rainy morning. Busily employed in the forenoon as yesterday. At noon the day became fine, when I then planted the few plants collected to bring home with to England. Afterwards went about three miles to the eastward for

Waikeea River in Byron Bay. From Maria Callcott's *Voyage of H.M.S.* Blonde *to the Sandwich Islands, in the years 1824-1825. Captain the Right Hon. Lord Byron, commander*. "Engraved from a Drawing made on the spot by Robert Dampier."

Waterfall in Byron Bay. From Maria Callcott's *Voyage of H. M. S.* Blonde *to the Sandwich Islands, in the years 1824-1825. Captain the Right Hon. Lord Byron, commander.* "Engraved from a Drawing made on the spot by Robert Dampier."

to get more young *Metrosideros* plants. In the evening heavy rain. Thermometer at 6am 73 at 12 80 & at 7pm 78.

5. A fine morning, with some light showers of rain till noon. Employed all day among my specimens & seeds &c. In the afternoon the party gone on Sunday to the volcano returned. This is my last night on shore at Byron's Bay, as I am requested to be on board tomorrow forenoon with all my traps, three days sooner that I wished, which prevents me from having the opportunity of a few more journeys in the wood above the harbour to great advantage. Thermometer at 5am 74 at 12 89 & 7pm 84.

6. A fine morning. In the forenoon packed up everything to go on board, & afterwards took them to the sea beach, when at four in the afternoon a boat came for me to go on board, once more to live in midst of bustle & noise the ship being crowded with natives & their provisions for Woahoo, besides several new canoes & 146 large bundles of tapa cloths, received by Kaumanna for rent. Thermometer at 6am 72 at 12 80 & 7pm 74.

7. A fine morning with light wind from S.W. weighed anchor at 5am & made sail for the island of Woahoo taking on board both the Queens we had brought from their, also two of the missionaries with their wives & families, besides upwards of 100 natives. At 7 again came to anchor for want of wind to get out [to] sea & a hour after got underweigh with the surveyor's schooner in company who had Mr. Young on board of her to be landed on the coast at where he lives on her way round to Karakaa Kua bay on the south side of Owhyee where Captain Cook was killed. At 2pm still light variable winds, lowered the boats to tow the ship [off] the land. At dusk fresh breezes & fine. Thermometer at 6am 71 at 12 76 & at 7pm 75.

8. A fine morning with fresh breezes from East. At 5pm made Diamond point near Hana-rura at Woahoo & at 7 reefed topsails, to shorten sail for the stand off & on the land for the night. Thermometer at 6am 73 at 12 78 & at 7pm 75.

9. A fine morning with moderate breezes from N.E. At 7am came to anchor [off] the harbour of Hana-rura, when shortly after the pilot (Adams) came on board & gave information that the *Asia 64* a Spanish ship with two brigs in company who were pirates on the coast of California was reported to touch here for supplies, by a schooner lately arrived from the coast. In consequence

of which all was soon bustle to get the natives on shore with the luggage as soon as possible that the guns might be all mounted in readiness for action. At nine o'clock His Lordship requested me to go on shore & get my plant boxes on board without delay as he did not know but he might perhaps get underweigh this evening for to join the schooner at Karakaa Kua Bay. At 4pm returned on board & found the natives with then gone & all the guns mounted & loaded, & the sailors in high spirits wishing for their reported enemy to come in sight to give them battle that they might gain prize. Thermometer at 6am 73 at 12 78 & at 7pm 76.

10. (Sunday) A fine morning with fresh wind from N.E. In the forenoon had church service performed at the usual hour. After dinner went on shore being informed we did not intend to sail before tomorrow forenoon, not to return here again. Therefore thought it necessary to go & see before I left for good what plants there was alive of those brought from England, which on my getting to Mr. Pitt's garden where they had been planted I met with His Lordship [&] Marin the Spaniard, whom I have before had occasion to mention, who both accompanied me to count their number, when we found them only six & without leaves. The most of the others Marin informed us, he had removed to his farm in the mountains, & were many of them doing well. To Mr. Marin for his diligence as a farmer the Sandwich Islands particularly Woahoo where he resides may be said to be under some obligation for his introducing & multiplying various kinds of our animals, which in a few years will be plentiful on the island. When I was about to leave Mr. Pitt I requested the favor of Marin to enquire of him if he would give me some patterns of their native tapa cloths for the Society & received for answer that he would consider of it. Remained on shore with one of the gentlemen on board at the Inn of Novara. Thermometer at 6am 76 at 12 78 & at 7pm 75.

11. A fine morning with strong wind, which raised clouds of dust in the town, as to be almost impossible for persons to walk along the different streets without being in danger of getting blind & nearly suffocated. During last evening I learned that Pitt had married since we had been at Owhyee a young woman who with her sister had before lived with him many years as his wives. Pitt is said to be about the age of 60 & his bride 19, their marriage is the second which has ever yet taken place on these islands. At one pm went on board & found all preparations made for getting underweigh when His Lordship came on board. Thermometer at 6am 73 at 12 76 & at 7pm 75.

12. A fine morning with fresh breezes from N.E. In the forenoon His Lordship on leaving the shore to embark was saluted with 13 guns from the forts, & was accompanied by Pitt, Boki & several other chiefs to take farewell of us. It was now generally understood by all on board that we were to leave the islands as soon as possible for Otahite after touching at Karakaa Kua bay for to take on board the surveyors out of the Sandwich island schooner given them for their use by the government to survey on the coast. Pitt while on board scarcely ever indulged himself with his accustomed habit of being seated, when seen in his hut on shore, but walked alone or in company, eagerly examining all parts of the ship till we got underweigh, when to our surprise he requested the favour of His Lordship to make all sail & get out for some distance to sea that he might see how we sailed & also wished to see the ship's company, who to gratify him were immediately mustered on the quarter deck with the band playing music all the time, when he was going round with His Lordship to examine the men & see their number who as is customary with them on board a man-o-war touched their hats as they passed them, which afforded to some of the young gentlemen rather severe remarks to be made upon him, for his having the honour of the sailors touching their hats to a half civilized savage, unaccustomed to such honours from his own bareheaded countrymen. At three pm we made in again for the land & sent them all on shore, when Pitt on leaving the ship was honoured with a salute of 13 guns from the *Blonde*, then afterwards made sail for Karakaa Kua bay with fresh breezes & cloudy weather which continued the same till sun-set. Thermometer at 6am 73 at 12 79 & at 7pm 76.

13. A fine morning with moderate wind from N.E. At 10 punished several of the men chiefly for drunkenness & disobedience of orders. At dusk wind still moderate & cloudy. Thermometer at 6am 78 at 12 80 & at 7pm 78.

14. A fine morning with light wind from West. At day-light saw the high land of Owhyee far above the clouds covered with snow. At 10am came to anchor in Karakaa Kua Bay in 26 fathoms & at one pm went on shore where I found several of the midshipmen who had gone before me in different canoes from alongside the ship with natives in them trafficking with persons on board chiefly shells of no variety. Some eating with coco-nuts that grew on the tree where we was shown a cannon shot had gone through the trunk, near where Captain Cook was killed, while others were close to the water's edge looking at the place where he fell, which is at times overflown by the tide of high water. When we had satisfied our curiosity here we next went to examine

several caves close at hand in front of a steep volcanic rock, which divides Karona from Karakaa Kua. The first of these caves that we went into was full of several dozens of old musket apparently the manufacture of various countries & here neglected probably for many years as scarcely any was now complete, where we saw them heaped up broken in pieces covered with rust. While searching the corners for curiosities two of the party found an old fishing rod & a club the former chiefly made of human bone the other of wood in shape of the thigh bone. The fishing rod was given to me, as a difficulty appeared to venture in taking it away without asking the chief who had power to give it, which trouble I willingly took upon me by getting it for myself, but was on seeing the chief at his hut, prevented from having it by the interference of His Lordship, who unluckily for me happened to be at the time present & who requested the club as well as the fishing rod to be restored to go where it was taken from. I felt as may be imagined not a little disappointed for his interference in so trifling a thing, that I was sure if the chaplain or any other of the [gun worn?] officers who since we had been at the islands were considered from their rank on board to be chiefs of some consequence like themselves, had wished for it they would not have received no denial. After the affair of the fishing rod & the club with the chief (Mai) & His Lordship we next directed our course to the spot where Captain Cook's body was said to be divided by the chiefs about half a mile above Karona on the top of a hill, when on coming to it we found the place had been held sacred for some time from the appearance of the old walls raised round of lava about 20 feet square & from four to five feet high & nearly the same in thickness. In the evening we reached across the other side of the bay at Karakaa Kua which like Karona is only composed of a few straggling huts of uneven broken pieces of beach lava situated near the sea side within places a few coco-nut trees, growing as it may appear astonishingly entirely upon lava without any signs of anything nutritious to support them. At Karona we noticed that the missionaries [had] a tolerably good meeting hut, not far from their provider Mai's hut & who is one of the most civilized & religious chiefs on the islands. Near where the immortal Cook was cut up to be divided among the chiefs, grow a fine white flowering plant of a low shrubby nature not met with in any other part before which belongs to the tribe *Diadelpha* & generally met with only on the lava near the sea coast, with scarcely any other vegetation no where near it. Plenty of seeds which I gathered I am in hopes will grow in England. Returned on board at dusk in a native canoe for want of a ship's boat being on shore on this side of the Bay. Thermometer at 6am 78 at 12 80 & at 7pm 79.

Missionary Houses in Byron Bay. From Maria Callcott's *Voyage of H. M. S.* Blonde *to the Sandwich Islands, in the years 1824-1825. Captain the Right Hon. Lord Byron, commander*. "Engraved from a Drawing made on the spot by Robert Dampier."

15. A fine day. Remained on board till the afternoon as I had given up thoughts of going to the woods to collect from the uncertainty of our stay, therefore could only employ my time picking up the few plants growing on the barren fragments of black lava, which lay in large masses occupying the sea coast for several miles in the interior, behind which the ground rise[s] gradually towards the woody mountains, when more vegetation appears, with cultivated patches of sweet potatoes, tarro &c. & a number of bread fruit & banana trees which give the rising ground here a few miles from the coast a luxuriant appearance. The whole coast on this side the island presents a prospect of the most horrid & dreary kind. The ground is everywhere covered with cinders & intersected in places with black streaks & hollow parts underneath which mark the course of lava from the mountain of Mouna Roa. In many parts the lava is piled irregularly on one another terminating in sharp points, as we before noticed, when at the volcanoes on the other side of the mountain. It is surprising to see the industry of the natives where they have taken to cultivate spots of ground about three miles from the coast. These they have cleared away the lava frequently to the depth of four feet to get to a more fertile soil of rich ashes & a light mould to grow sweet potatoes, tarro & other vegetables, with the tapa which are generally enclosed with stone walls or rather lava cleared off the ground. The tarro is but little cultivated anywhere on Owhyee in ponds of water, as we noticed at Woahoo & Mowee & is in consequence of much smaller growth, but it is more dry & farinaceous & equally good to that grown in water. The sea here like the other parts of the islands we touched at abound with a variety of excellent fish many of whom possess handsome colours. The fertile appearance seen at Byron's Bay on the opposite side of the island is here changed to a more gloomy & inhospitable dreary appearance. What little water is met with on the coast in hollow places among the lava is generally brackish by the salt water. At Karakaa Kua beside a sandy beach we was shown a large heap of stones where Captain Cook had his observatory which was formerly a morai & religious place for human sacrifices. We also noticed near the same place three coco-nut trees with cannon shot holes through them from the firing which took place upon the natives when Cook was killed. In the little gardens which surround the native huts who are fenced with walls of broken lava, grow scattered on a stony surface a few plants of the tobacco & banana trees for immediate use. During the time we remained on shore we had the opportunity of seeing a dog getting cooked for one of the chiefs who consider them better eating than any other of their animals, which is done in the same manner with cooking hogs. The

throat is first cut & when the bleeding ceases they then place the animal over a large fire to get the body deprived of the hair, instead of using hot water, & when the inside is cleaned, it is afterwards filled with hot stones & buried in the ground for some length of time covered over with leaves & earth. Their filthy manner of killing dogs, which is the employ of severals at the time & are immediately after cooked & eat by the chiefs, is too disgusting to describe, particularly the ravenous way they consume the entrails, while the rest is cooking. When I returned on board at dusk I found His Lordship with several others had been at the morai about nine miles to the eastward of Karakaa Kua Bay where they brought away from several idols that had been formerly worshipped by the natives & some other curiosities. Having a wish to see this morai before we left, I asked permission of His Lordship who readily granted me, my request to go with another party made up to see it tomorrow. Thermometer at 6am 76 at 12 81 & at 7pm 78.

16. A fine day. At 10am went in company with a party from on board to see the morai, or temple formerly the sacred depository for the remains of the high chiefs. On our way we met the old priest in his canoe coming on board, who is alone entrusted to enter the morai, whom we took back with us in our boat, which on getting to the place we found it was situated on the east point of a small bay pleasantly surrounded with several plants underneath thinly scattered coco-nut trees, but like Karona & Karakaa Kua, having no signs of cultivation about its neighbourhood. When we was about to enter the morai the old priest who had on a straw hat & a cotton shirt now took them both off and only left on his maro. On opening the door which was made of a few old boards, we on entering found only an empty hut of the most filthy description, with a quantity of human bones in heaps under mat at each end of the hut, not yet many of them dry & disgusting to the sight, in the middle were several effigies in representation of the deceased chiefs, tied to a bundle of tapa cloth containing the bones of each person whom the effigies represented. Most of the effigies were made of wood, but the one representing the late Tamahamaah was substituted by a mask of European manufacture & was particularly more finely dressed than the others with handsome native tapa cloths. The party who had been here before us yesterday, having carried away mostly every thing which could be given them, left the inside now for us entirely empty with only bones & skulls, which we durst not venture to ask for, & to come away without some memorial from the only morai now standing on these islands we did not feel alto-

Morai Near Karakakooa. From Maria Callcott's *Voyage of H. M. S.* Blonde *to the Sandwich Islands, in the years 1824-1825. Captain the Right Hon. Lord Byron, commander*. "Engraved from a Drawing made on the spot by Robert Dampier."

gether included, we therefore got the permission of the old priest to take away some of the ancient weather beaten carved figures outside which on getting them into the boat, two of our party accompanied me home by land, while the rest returned on board in the boat. The morai is a small thatched hut fenced round with sticks to the height of about six feet kept together by two rows of [lashing] bars, where in front at the entrance stand upright fixed in the ground or rather lava several various sized wooden rude carved hideous figures in representation of their former gods which they now set but little value upon & are rarely met with in the huts of the natives. While we travelled home by the sea coast on uneven masses of lava, I met with only a few plants, such as tufts of low growing *Euphorbias, Convolvuluses, Sidas*, but mostly the diadelphous plant seen common near the spot where the chiefs shared between them the body of Captain Cook. Also a species of *Hibiscus* with yellow flowers, which is much used here by the natives to dye yellow their tapa cloths. About half way between the morai & Karakaa Kua we passed a number of small heaps of lava raised over the dead slain here at one of Tamahamaah's first battles for the conquest of the island. At four in the afternoon returned on board when we found the old priest waiting for us to make him some present for the images he had given us at the morai, for which we gave him a few articles of clothing & seemed perfectly satisfied more so than if we had given him money. Having been but a short time on board when I was sent for by the commanding gentleman at the time, for me to take away my wooden gods, who he said were only for the cook to boil the coppers in the galley, forgetting himself that he was more lucky on visiting the morai before me to get feather images of more consequence besides other curiosities that would be denied me like the fishing rod, because I was not considered by the natives to be a chief like himself & others on board. Thermometer at 6am 76 at 12 79 & at 7pm 76.

17. (Sunday) A fine day. Went on shore in the afternoon to see after my linen given to a black American to be washed soon after we had come to anchor, who yet I found had left them untouched in the bag as they were given him, which put me out of temper with him for being disappointed to go to sea again without having them washed. When I had sent them on board for fear of the natives stealing them, I then accompanied several of the gentlemen from on board going to see the monument erected yesterday to the memory of Captain Cook by His Lordship on the top of the black lava hill above Karona where it is said his body was divided by the chiefs. This monument

now erected is simply a capstan bar from the ship painted white fixed in a heap of lava that had been raised by the natives formerly when they made the square wall round this spot, whom they now for years look upon as consecrated ground. This bar stands about sixteen feet high with a small copper plate at top, having the following inscription upon it in small print rather difficult to be read on account of its height—

In memory of Capt. James Cook R.N.
who discovered these islands
in
The year of our lord
1778
This humble Monument is erected
by his fellow countrymen in the
year of our lord 1825.

When I returned on board in the evening I found preparations were making to get underweigh when the land breeze blew off the shore to enable us to get out to sea. Thermometer at 6am 76 at 12 79 & at 7pm 78.

18. A fine morning. At 12:10am, weighed anchor & made sail with a light breezes from W. N. W. for Otahite, now leaving these interesting islands, having on board all kinds of provisions and about one hundred live hogs. Yet for all the civilities & plentiful supply of provisions received from the natives during our short stay of little more than two months, none on board I believe regretted leaving so fine a climate, or the civilized beings who inhabit it, which with the exception of a few in high rank, who have adopted a little the European costume, still retain their manners & customs nearly the same as given by the unfortunate Cook who fell a sacrifice to their barbarity. The dagger which is said to have committed the fatal deed is I am now informed in the possession of a gentleman on board with every other particular circumstance relating the misunderstanding which took place between the parties at the time. But I am inclined to doubt whether such a thing as a dagger at that period had been among them, or why should it be held secret from Vancouver & other navigators who were equally anxious to gain every information respecting the occurrence that took place at the time & with what weapon the fatal act was committed. The small English drum presented by Cook to the King is now in His Lordship's possession with some other things presented by him when he

visited the islands, which have been in the morai for many years, & were given His Lordship with other interesting things at the time he visited that place a few days ago. Thermometer at 6am 76 at 12 81 & at 7pm 78.

The following are a few names of their plants.

NATIVE NAME	ENGLISH NAME
Ana or ava	*Piper* the roots of which they chew to intoxicate themselves
Ape	large species of tarro
Cueni	[*Aleurites*] or candlenut
Corena	Indian corn
Halla	Pineapple, not plentiful
Hapa-houri	Water & musk melons
Ipo	Gourd for calabashes
Raikai	Onions or shallots
Rassoeho	Cabbage
Rouh	*Acacia*, which they make canvas
Marya	Banana
Niu	Coco-nut
Aicoi	*Capsicome*
Ohia lowholla [lehua]	*Metrosideros*
Ohia	*Eugenea*, or Rose Apple
Owawella	Sweet potatoes
Otho	Sugar
Paka	Tobacco
Pihu	Fish
Passo	Kidney Beans
Pihah	Arrow root. *Tacia* species
Poolo poolo	Cotton tree
Roio	Sugar Cane
Rilu	Bread fruit
Tarro	*Alum exulentum*
Tapa	*Broussettia* species. Bark make cloth

19. A fine morning with fresh breezes from E. N. E. At pm squally weather. At dusk moderate & cloudy. Thermometer at 6am 76 at 12 82 & at 7pm 76.

20. A fine morning with fresh breezes from E. N. E. At 4pm squally with rain. At dusk no alteration in the weather. Thermometer at 6am 78 at 12 81 & at 7pm 78.

21. A fine morning with light airs N. E. At dusk moderate wind which shifted to E. N. E. Thermometer at 6am 78 at 12 80 & at 7pm 78.

22. A fine morning with light breezes from E. N. E. at noon fresh breezes & cloudy. At [dusk] squally with rain. Thermometer at 6am 78 at 12 82 & at 7pm 79.

23. A fine morning with light airs from E. N. E. At dusk moderate wind, which shifted to S.E. Thermometer at 6am 80 at 12 82 & at 7pm 82.

24. A fine morning with light variable winds. At noon calm. At dusk moderate breezes from E. S. E. & fine. Thermometer at 6am 81 at 12 83 & at 6pm 80.

25. A fine morning with light airs from S.E. & at dusk moist with rain. Thermometer at 6am 81 at 12 83 & at 6pm 80.

26. A fine morning with light air from S.E. at dusk moderate with rain. Thermometer at 6am 81 at 12 83 & at 6pm 80.

27. A fine morning with fresh breezes from E. by South. At pm crossed the Equator. At dusk fresh breezes from E.N.E. Thermometer at 6am 79 at 12 81 & at 6pm 80.

28. A fine morning with fresh breezes from N. E. by East. At dusk wind shifted to E. N. E. moderate and fine. Thermometer at 6am 78 at 12 80 & at 6pm 78.

29. A fine morning with light airs from E. N. E. at noon the wind shifted E. S. E. & at dusk again to E. S. E. moderate & cloudy. Thermometer at 6am 86 at 12 81 & at 6pm 80.

30. A foggy morning with variable wind & some light showers of rain. At 7am saw low land from the mast head bearing S. by E. At noon shortened sail & lowered the whale boat & cutter to try landing, when both the surveyor

& the naturalist went in the whale boat, & two of the Lieutenants, surgeon & chaplain with myself in the other, which on our getting near the shore saw numbers of sharks & other fish round our boats, also flocks of sea birds hovering over our heads, till after landed through a high surf on a fine coral sandy beach, covered with great quantities of shells, most of which were of the cockle kind & very large weighing upwards of fifteen pounds. By the sea side & other places on shore we noticed great numbers of small red crabs that occupied the shell of another fish. When we had landed we afterwards divided in separate parties to examine the island, which we found was low & flat covered mostly with vegetation, but without variety, being such as sidas with two or three kinds of short grasses & in places round the outside by the coast a few low bushes of a species of *Tournefortia*, who are seen covered with birds that were remarkably tame & suffered us to approach them to be easily taken with our hands. During our short stay we met with several remains of huts, both on the west & south east side that had been built of coral slabs without any timber nearly of a square form, which on our falling in with them gave us every reason to suspect that the island had been formerly inhabited by some persons or other, who had been probably unfortunately shipwrecked & suffered to remain which they lived without being discovered. After we found out the remains of these huts at one of which the surveyor left a bottle sealed up that he had brought with him on shore for that purpose containing some written inscription which I am ignorant of its contents, I endeavoured to find out signs when there might have been places cultivated, but without success, & only found that their was some little fresh water in holes among the coral rocks at a good distance from the sea side. At the moment our stay on shore became most interesting we had the mortification to hear the report of the recall gun fired for us to return on board, which prevented our making any farther search that might enable us to trace out some knowledge of the persons who had lived on the island, which very probably had our stay been longer to make better search some important information would have been made. Perhaps not unlikely something concerning the unfortunate Mr. de la Perouse & his crew, who the world still remain doubtful where he was lost in these seas. This little island which is a new discovery of the *Blonde* & since named Malden island in honour of our surveyor is about twelve miles in circumference & very low scarcely seen fifteen miles distant out at sea & lies in Latitude [ ] & Longitude [ ]. It abounds with sea birds of various kind, but without any quadruped animal excepting a small brown species of short tailed rat, which seemed so wild that on our approach they got into

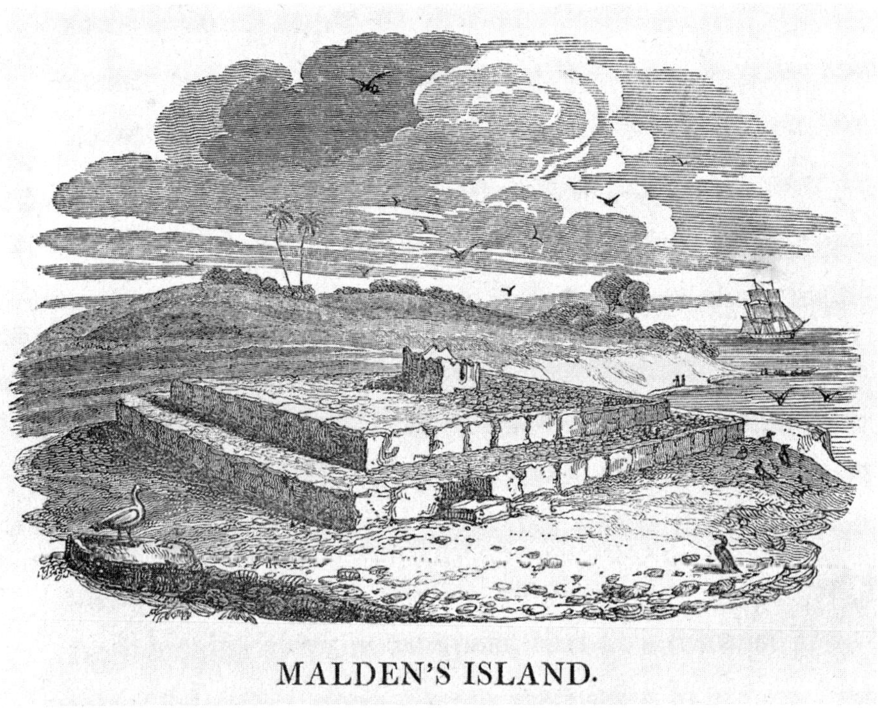

Malden's Island. From Maria Callcott's *Voyage of H. M. S.* Blonde *to the Sandwich Islands, in the years 1824-1825. Captain the Right Hon. Lord Byron, commander.*

their holes & never could get any of them. As we returned on board, we again made sail with strong breezes & at times squally with rain. Thermometer at 6am 79 at 12 81 & at 6pm 80.

31. A fine morning with light winds from S. E. & times some light showers of rain. At noon & the evening, no alteration. Thermometer at 6am 79 at 12 82 & at 6pm 81.

EDITOR'S NOTE: After leaving the Hawaiian Islands, the *Blonde* sails back to Valparaiso, Chile and eventually arrives in Britain on March 15, 1826.

PART II.

EXTRACT OF JAMES MACRAE'S DIARY,

1922

With Lord Byron at the Sandwich Islands in 1825

Being Extracts from the MS Diary of James Macrae, Scottish Botanist.

HONOLULU
1922

With Lord Byron
at the
Sandwich Islands
in 1825

Being Extracts from
the MS Diary of James
Macrae, Scottish Botanist

Originally published in
Honolulu
1922

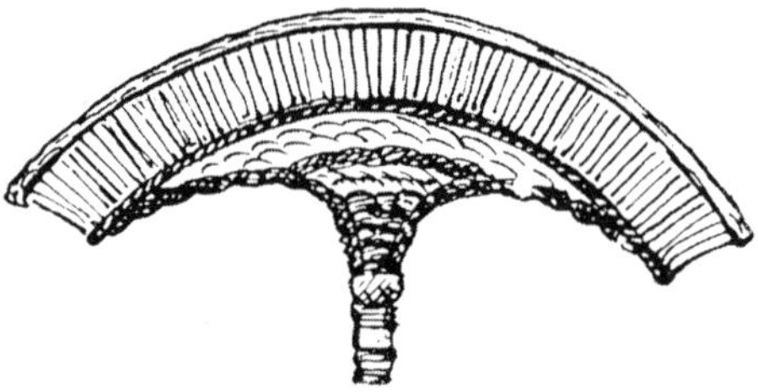

Old Style Hawaiian Fan

FOREWORD

KING LIHOLIHO IOLANI (Kamehameha II.) and his queen, Kamamalu, died in London while on a visit to see their "great and good friend" George IV. In order to show its respect for the royal pair who had travelled so far and who had come to such an untimely end, the British government sent their remains back to Honolulu in the *Blonde* frigate, commanded by George Anson, seventh Lord Byron, who happened to be first cousin to the poet and successor to his title, and who was also a grandson of Admiral Byron of the *Dolphin*, one of the British explorers in the Pacific previous to Captain Cook.

The Horticultural Society of London (now the Royal Horticultural Society), hearing of the intended departure of the *Blonde* for the Sandwich Islands, and wishing to help the natives of that group, obtained permission to send on the *Blonde* a fine collection of plants, considered suitable for the climate of "Owhyee" under the charge of James Macrae, a young Scotsman, trained as a plant collector and horticulturist. The plants were to be distributed as a gift among the chiefs of the islands. By the same opportunity, John Wilkinson, a skilled agriculturist, was induced, evidently through promises held out to him by Boki, one of the chiefs who had come to London in the train of the deceased King Liholiho, to come out to Honolulu with the intention of starting some kind of tropical farm on land to be given him by Boki.

Hitherto, the only published account of the voyage of the *Blonde* is the work compiled by Mrs. Maria Graham, and issued in London in 1826. This work is supposed to have been based on materials supplied by the diary kept by Mr. Bloxam, the chaplain to the *Blonde*, supplemented by details taken from the diary or journal of one of the midshipmen on board the same vessel. The writer of "An Examination of Charges against the American Missionaries at the Sandwich Islands as alleged in the Voyage of the ship *Blonde* and in the

London Quarterly Review," published at Cambridge (U.S.A.) 1827, has a very poor opinion of this account of the voyage as set forth by Mrs. Graham, and terms it "nearly worthless."

During the course of the voyage of the *Blonde* to and from the Sandwich Islands, which included visits to various ports en route, James Macrae, when he had the chance, was a diligent collector of plants and objects of natural history. On returning to England in 1826, his herbarium and diary were handed over to his patron the Horticultural Society. The herbarium was probably delivered to Kew or the British Museum in order that the various plants collected by Macrae might be scientifically catalogued and described. His manuscript diary, however, has reposed, until this date, on the shelves of the library of the Royal Horticultural Society. The present writer thought it would be worth while to have that portion of Macrae's journal which refers to the Sandwich or Hawaiian group made known to the public. He has therefore obtained the consent of the Council of the Royal Horticultural Society to its being published.

To those who are interested in the early history of the Sandwich, or as they are now termed, the Hawaiian Islands, it is believed that in Macrae's journal will be found some curious glimpses of men and manners as they existed in the islands at the close of the first quarter of the nineteenth century. The Hawaiian chiefs and people were then beginning to come in contact with the haole or white man, but were still masters in their own land. The people grew their own taro and caught their own fish, and did not rely on Asiatics to do it for them. They were still living in grass houses built by themselves and they were clad in the native tapa or kapa cloth.

Like his fellow countrymen, Doctor Archibald Menzies and David Douglas, who had visited and explored the Hawaiian Islands, James Macrae was a botanist and mountain climber, and the account of his visits to Mauna Kea and Kilauea volcano and his rambles on the island of Oahu are worth preserving.

After his return to England, Macrae was appointed by the British Colonial office to the post of superintendent of what was then called the Colonial Gardens at Ceylon. Whether or not these gardens were at Colombo or at what is now called the Royal Botanic Gardens at Peradenyia, near Kandy, the present writer has, so far, been unable to find out. Neither has he been able to ascertain anything further about the career of Macrae. As Macrae was a collector and outdoor man, botanical writers have little to say about Macrae. Had he been a "systematist," a party such as Linnaeus or de Candolle, who sat at home at ease and catalogued the plants that had been gathered by hardy explorers like Macrae, then no doubt we would have known more about the facts of his life.

In the botanical world Macrae is remembered by Macraea, a tribe of Chilean plants belonging to the order of *Geraniaceae*, and in the Hawaiian flora, several plants have Macrae for their trivial names. Some of these may be mentioned here, viz.:

		Native Name.
Asplenium Macraei	Hook. & Grev.	Pamoho
Coreopsis Macraei	Gray	
Cyathodes Macraei	D. C.	Maieli or Puakeawe
Cyrtandra Macraei	Gray	
Davallia Macraeana	Hook. & Arn.	Laukahi
Depavia Macraei	Hook. & Grev.	
Peperomia Macraeana	D. C.	Alaala wainui
Phyllostegia Macraei	Bentham	Kapana
Rubus Macraei	Gray	Akala
Vaccinium Macraenum	Klotzsch	Ohelo

In the *Index Kewensis Plantarum*, other plants named after Macrae may be found.

Mrs. Maria Graham, the compiler of "The Voyage of the *Blonde*," does not mention Macrae by name, but the following sentences in her work refer to him. "It is to be regretted that the practised collector of botanical specimens who went in the *Blonde* to the Sandwich Islands, should not have furnished any account of the plants, useful and curious, which he collected for the Horticultural Society, or that some of the very enlightened members of that society should not have done so. The field is in some respects new; and it is acknowledged by all the foreign navigators that the collection made during the *Blonde's* voyage is one of the most curious in Europe."

The undersigned has added a condensed log of the *Blonde* before reaching and after leaving the Sandwich Islands. He has also appended explanatory notes to Macrae's diary, and furnished several illustrations taken from engravings, photographs and autographs in his private collection.

WM. F. WILSON.

Honolulu, September 30, 1922.

NOTES ON VOYAGE H.M.S. *BLONDE* FROM WOOLWICH, LONDON, TO HILO, HAWAII

1824

September 8 — H.M.S. *Blonde* of 46 guns, commanded by Captain Lord Byron, then lying at Woolwich, received on board the bodies of the late King and Queen of the Sandwich Islands.

September 28 — The suite of Sandwich Islanders got on board at Spithead, and the ship sailed the same day.

October 18 — Reached Madeira and stayed there until 23rd October

November 27 — Reached Rio de Janeiro and left on December 18 for St. Catherines for the purpose of obtaining provisions and water.

December 24 — Reached St. Catherines and sailed again on January 1,

1825

February 4 — Anchored in Valparaiso Bay.

February 8 — Kapihe, "Admiral of the Sandwich Islands," and one of the late King's suite, died today from an attack of apoplexy. He was a skilled player at the Hawaiian game of draughts, called "konane."

March 16 — Reached Callao.

March 25 — Arrived at Galapagos Islands, and stayed there until April 2.

April 3 — Boki, who had already been baptized at the islands, partook of the Lord's Supper.

May 1 — Liliha and the other Sandwich Islanders on board were baptized by the chaplain of the *Blonde*. Lord Byron was their sponsor.

May 3 — Land sighted off the coast of Hilo, Island of Hawaii.

IOLANI
(King Kamehameha II.)
From a life sketch made in London

JAMES MACRAE'S DIARY

H.M.S. *Blonde* Arrives At Hilo, Hawaii

May 3, 1825. Made the S.E. point of Owhyee, and at 12 got close in with the land, which afforded us a pleasant contrast to the Galapagos Islands lately left. Here green slopes were seen beautifully interspersed with trees, and in many places along the coast, rows of coco-nuts with leaning bushy tops. The country in this part rose inland with a gentle ascent, and did not appear to us much intersected by deep ravines.

By 2 o'clock p.m. we hove to off Heddo Bay, and lowered two of the ship's boats, one for the surveyor and his assistant, and the other for the master to go in shore to try soundings in order to ascertain whether we could return here to refit the vessel after depositing the bodies of the King and Queen of the Sandwich Islands at the seat of government at Woahoo. While the boats were away two ragged white men came on board and reported that the ship was standing on not far from a reef which lay most way across the mouth of the bay. All was instantly bustle and noise to put the ship about and stand out to sea farther from the land.

We learned from these men that news had been brought them a month ago, by an English whaler from the port of London, of the King and Queen's[1] death. We were also informed that the natives of the island of Atooi to leeward, had rebelled two months ago against the Regent in consequence of the chief of that island having died since the king went to England, and that owing to the arrival from America some months past of the son and heir to the deceased chief, who claimed the rights of his father. Before he was defeated by the superior force of the Regent, who commanded in person, the rebellion occasioned the loss of upwards of two thousand natives, besides two Europeans. Yet we were assured that this same chief, after being captured, is at present suffered to go at large in the town of Hanarura at Woahoo.

We had not been long off the bay of Heddo when several of the natives visited us from the shore, paddling off in their canoes, and bringing with them sugar canes, taro (*arum spec.*), fish and hogs. The sugar canes were generally of the purple and striped kind and of a large size. The dress of the men consisted of the maro, a piece of various dyed cloth of their own manufacture, eight or ten inches broad, and about three yards long, which they pass between the legs and tie around the waist. The dress of the women consisted generally of a larger piece of cloth of similar manufacture, tied carelessly around the waist, forming a short petticoat reaching halfway down the thigh. This was the common dress of all, who came off to the ship, excepting that of the head chief, who is commonly honored with the title of governor, and that of one of his wives. Both were of more than ordinary size from overgrown

---

1. Liholiho and Kamamalu. The name Liholiho, or Rihoriho, is a contraction of Kalaninuiliholiho (the great black (?) heaven). On state occasions, Liholiho was induced to take the title of Kamehameha II, but his favorite name was Iolani (the great Io or god of heaven). Io was the supreme deity of Polynesia, and "his name was held to be so sacred that none but the priest might utter it at certain times and places" (Tregear). Io was recognized in New Zealand and at Tahiti and other islands in the South Seas. In Hindustan, the home of the Polynesian race, Io is known as Deo, Dyu or Dyo. In the ceremony of fire-walking or "umuti' as practised at Raiatea Island, the priest, before leading his followers across the fiery pit, invoked the aid of the supreme god Io. Although the word Io appears in Andrews' Hawaiian dictionary under eleven different headings, none of these mention that one meaning for the word is that of the great deity of Polynesia. Liholiho, having been initiated into the inner mysteries of the Hawaiian temple, must have understood the signification of the word, and hence his fondness for it as one of his own names. The inscription in poor Hawaiian, on the outer coffin made in England for the body of Liholiho, winds up as follows: "Aloka ino no komakoa Elii Jiolani," i.e., "Aloha ino no ko makou Alii Iolani," or in English, "Deep love to our chief Iolani."

Kamamalu is a contraction of Kamehamehamalu (the shade of Kamehameha), contracted to Kamehamalu and afterwards to Kamamalu. On the brass plate fixed to the coffin made for her in England, she is termed "Tamehamalu Eli," i.e., "Kamehamalu Alii."

corpulency. The governor's dress consisted of a dirty cotton shirt, a pair of blue China nankeen trousers and an old straw hat. That of his wife, the common petticoat used by the other females, with the addition of a piece of native made cloth tied round her neck, hanging loosely over her body and open in front like a cloak. Both these, on coming on board, touched noses, after the manner of salutation of their country, with most of the Sandwich Islanders we had brought out from England, and immediately after went with them below to their cabins, where they all soon commenced a disagreeable howling noise in lamentation of their late king and queen.

All the canoes of the natives that came alongside the ship had outriggers, the same as described by Capt. Cook and others. That in which the governor came on board was upwards of 40 feet long and very neatly made.

At five o'clock the boats returned, bringing information that the anchorage was good and the entrance to it sufficiently large for ships of the line.

Sails For Maui

We now made sail for the island of Mowee, where, his Lordship was informed, Queen Kanamanah[2] at present resided, to take her on board with any other people of consequence on that island who might wish to be present at the funeral, like the governor and his wife, who, with several others, had already come on board at Heddo. The neighborhood of Heddo Bay to the east, is thickly covered with wood from the seacoast everywhere back in the interior as far as we could see. The country is free from fog, and to the west much intersected with deep ravines ending by the sea in abrupt cliffs, down which fall several beautiful cascades of water.

The native huts appeared most numerous round the edge of the bay, being pleasantly situated under thick groves of coco-nut and bread fruit trees. Among the numerous trees that ornament this delightful spot, are numbers of the candlenut (*Aleurites*) so easily distinguished at a long distance by their white shining leaves. These grow everywhere on the coast.

May 4. Saw the summit of Mouna Kaah[3] towering far above the clouds, and in places covered with snow. We also saw Mauna Roa appearing not far off from the other, but without snow. The grand appearance of both these mountains as seen from on board the ship, upwards of 100 miles out at sea, would repay the traveller for coming so far on purpose to see them.

2. Kaahumanu.
3. Mauna Kea.

We had scarcely done looking at the mountains of Owhyee, when the highland of Mowee appeared in a similar way ahead, but nowhere covered with snow.

West Maui

In the course of the next hour or two, we were running along the shore under the high land, and we noticed that this side of the island was intersected with numerous deep ravines, down which fell several waterfalls, at no great distance from each other. These were admired by all on board, most of whom anxiously desired to be on shore that they might satisfy their thirst from such beautiful water. For the last month we had been on short allowance of this article, which was often served out not drinkable. The land on this side of Mowee rose to the peaks of the mountains much more steeply than anywhere noticed at Owyhee. Like that island it was well wooded, and along the coast covered with verdure, yet little cultivated, although the native huts appeared everywhere more numerous and at no great distance from each other, generally upon the ridges near the coast for the convenience of fishing.

Lahaina

When we got near enough to the shore we lowered a boat and landed all the Sandwich Island people brought from England. These were dressed in new mourning suits, which formed a rare contrast beside their naked countrymen. When they landed on the beach, they were received by some thousands of natives making a very disagreeable loud howling noise, ceasing at intervals for a few minutes, then commencing again as before, which they kept up in this manner most of the night. About seven, came to anchor in eleven fathoms on a coral bottom, not far from a reef with a high surf on it. During the night a native schooner was sent to Woahoo to inform Mr. Pitt[4] and Mr. Charlton[5] of our arrival.

May 5. A fine morning. Lowered the ship's boats for the purpose of getting a fresh supply of water. Got permission to go ashore, but promised not to go out of sight of the ship, it being uncertain when we might get under

4. William or Billy Pitt, the nickname given by foreigners to Kalanimoku or Kalaimoku, governor of the island of Oahu.
5. Richard Charlton, first British consul to the Sandwich Islands. His jurisdiction as consul extended to the Society and Friendly Islands. Died at Falmouth, England, on 25th December, 1852.

weigh. On going ashore I met three poor ragged white men at the watering place, who offered their services to accompany me through the native huts and plantations. To a stranger it was difficult to find the nearest way, from the numerous taro ponds almost everywhere among the huts. The taro, I noticed, was cultivated to a greater extent than any other esculent. The ponds of water in which they cultivated it, are commonly square and of different sizes, about four feet deep from the level surface, and the bottom covered with water to a depth of from 2 to 2 1/2 feet. The water is constantly supplied from a neighboring stream by means of a small canal emptying itself from one pond to another, until at last it reaches the seashore. These ponds are used for keeping fish and ducks as well as for growing taro. In general they are kept free from weeds and rushes, and are planted with surprising regularity, not far behind what one would expect from a more civilized race of people. The singular method which they have adopted of growing this farinaceous esculent always in water has originated beyond doubt at an early period with themselves. Perhaps it was from their extreme fondness for that element which they so often frequent during the day, in a manner similar to the feathered aquatic tribe. When they have occasion to weed these ponds, they are generally up in mud and water above the knees, which to any other race not so much accustomed to water, would be found inconvenient. They, as far as I could learn, have no particular season for planting or for taking up the taro, but go on using the largest roots as they occasionally want them for food from the ponds that are most forward till they become nearly done. Then they turn the water off and drain the pond dry, placing whatever fish it may contain into the nearest pond. The old ponds from which the taro has been harvested are left for a few days exposed to the sun for the mud to harden. Then round or square clumps are thrown up about two feet high and seven or eight in circumference, at short distances from each other, always in straight lines, either across or lengthwise. In these clumps are planted the crowns of the taro, from eight to twelve in number alternately about a foot apart. The crowns which they choose for planting are the top parts of the taro cut across about half an inch, adhering to the leaves, which they shorten to the length of eight or nine inches.

Round the sides of the ponds, by the edge of the water, is often planted what they term the tea tree[6] (*Dracaena terminalis*), the root of which they cook underground as is customary with them in everything else that requires cooking before it is eaten. This they afterwards pound and put into vessels

6. In Hawaiian "Ti" or "Ki."

full of water and let it remain there for a few days to ferment, when it makes a good substitute for beer of a very intoxicating quality. A piece of the cooked root which was brought to me to taste was as sweet as sugar cane, and in my opinion contained an equal quantity of saccharine matter. If the common process generally used for the making of sugar was here once to be introduced, it might be turned to the same advantage as sugar cane for manufacturing sugar. But whilst there is no emulation among the natives for commerce or cultivating more than is necessary to supply their present wants, the value that might likely be found out from this plant with so little trouble will probably yet remain for a length of time imperfectly known.

One of the said white men took me to his little garden which surrounded his hut. Here I noticed *Cytisus cajan* or the common pigeon pea of the West Indies, which he said was given him last year by the captain of an American whaler for coffee, together with some seeds of the lima bean (*Phaseolus lunatus*), both of which were now in bearing for some time. But being as yet unacquainted with the way of using them for food, he always felt afraid to touch them for that purpose. These, with a few light-red coloured cabbages, two sorts of sweet potatoes (the red and white), water melons, pumpkins, a few patches of sugar cane, two or three ponds of taro and some bitter gourds that are used by the natives for various purposes, such as calabashes, were, with a hog and a few fowls, all that he had for the support of himself and family.

His wife is a native woman, by whom he has had three children. He has often applied, without success, to the missionaries to baptize his children, but they are considered by them to be born out of wedlock.

Beyond the huts and plantations, I observed but few plants. Some were Cleome, Argemone, and two sorts of Malvas, three of Sidas[7] (two of which the natives use for wreaths and necklaces, by stringing the flowers on a thread made from the bark of the tapa plant). In the cultivated grounds of the natives I noticed they had no plantains and only three sorts of bananas. One of these was much shorter than the others, and different to any I had ever seen before. The red banana[8] common in Otahiti, has not yet reached

---

7. The ilima of Hawaiians, until recent years very much used by them in making wreaths ("leis") for the head or neck. Ilima blossoms have given place to meaningless colored paper wreaths.
8. Fei (*Musa fei* or *Musa uranascopus*), the wild mountain banana of Tahiti, which forms a considerable part of the daily menu of the natives of that island. The large bunch of fruit grows upright from the top of the stem instead of hanging down like the ordinary kind of banana. It requires to be cooked before being eaten. In Hillebrand's *Flora*, page 434, it is stated that the fei had been introduced into Oahu from Tahiti, and that it grows in a few of the higher ravines of Oahu. Is this a fact? Seemann states that this species of banana occurs in Fiji.

the Sandwich Islands, but will now, no doubt, soon be introduced by Mr. Charlton, who is to visit that Island once every year.

Description of Lahaina in 1825

The town of Lahaina on Mowee Island, is composed of a number of low thatched huts, scattered along the sea shore for about a mile in length, and in places nearly half that distance in breadth. It lies on a level flat at the foot of the high mountains which rise abruptly in the central part of the island. At the west end of the town is a small grove of coco-nut trees. More towards the town on the sea beach, is a mud battery in bad repair, mounted with five small cannon, in the same neglected state. Within its walls is a small mud hut, whitewashed outside, where had been buried lately one of Tamahamaah's queens.[9] Near this fort, the missionaries have a small thatched chapel, with dwelling houses and garden grounds. Here I was shown the only grape vine on the island. It was yet but young and never produced fruit.[10] Close to the beach, nearly in front of the town, stands a brick house of two low stories, whitewashed outside, built sixteen years ago by Tamahamaah for his favorite queen Kaumanna, which she never inhabited, choosing rather to live, after the native fashion, in a thatched hut close beside the other.

The town has no regular streets, being all cultivated and rather difficult to get from one end to the other on account of the taro ponds. It looked like a well cultivated garden, divided into allotments by mud walls enclosing each family hut and garden. Sugar canes grow with little trouble on the narrow ridges between the taro ponds, where they have also at times, cotton, tobacco, and cabbages. The tapa tree (*Brousonettia*), from the bark of which they manufacture the cloth they wear, occupies a large proportion of their ground. It is neatly planted out in rows and kept free from weeds.

I was informed that the number of natives living at Lahaina exceeded 6000, and some years before had been far above that number, but since then hundreds had died in a short time from some unknown fever.

At four p.m. the ship's boats having brought off the necessary water and provisions, all on shore were ordered to return on board at once. At six we weighed anchor and made sail for the island of Woahoo, having on board

9. Keopuolani, the bluest-blooded wife of Kamehameha I. She was mother of Liholiho (Kamehameha II.) and Kauikeaouli (Kamehameha III).

10. After the date of Macrae's visit, Lahaina became celebrated for its grapes, and whenever an inter-island schooner or steamer touched there, shore boats came off laden with grapes, mangos, etc., for sale. This custom exists no longer.

about forty natives, among whom were the mother of the deceased queen, and her daughter, an interesting young girl of ten,[11] sister to the present youth proclaimed king. This princess was dressed in deep mourning brought her from England by Madame Boki. Several of the chiefs who came on board were also clothed in European fashion, while others of inferior rank had only a shirt and maro without a hat.

Arrives At Honolulu

May 6. Fine morning. Made the island of Woahoo and at 7 a.m. got under Diamond Head and saw the Bay of Wytitte covered with thick groves of coco-nut trees, where lay underneath many of the native huts. Behind these appeared the woody mountains rising in places abruptly and intersected by numerous valleys having on each side high peaked ridges covered with green forests.

At 8 we were off the harbor of Hanarura and made signal for the pilot. Shortly after, Captain Charlton came on board to conduct us to safe anchorage outside a coral reef which runs across the mouth of the harbour, nearly a mile from shore, and having only in one part opposite the town a narrow passage over a bar from 3 to 4 fathoms of water. This was thought too shallow for our ship going over to a more convenient place in the inner harbour. At 10 on coming to anchor in 20 fathoms we fired a salute of 15 guns, which was, to our surprise, returned with the same number from both forts on shore, where we noticed was hoisted the national flag,[12] having the Union of Britain in the upper corner.

---

11. Nahienaena.
12. It is generally believed that the Hawaiian national flag was designed by a Scotsman, Captain Alexander Adams, who arrived at the islands in 1809, and resided there until his death in Honolulu on October 27th, 1871, at the advanced age of 91 years. He lies buried in Nuuanu Cemetery alongside his old crony, Andrew Auld. The inscription on their common tombstone reads as follows:

Sacred to the memory of

Alexander Adams a native of Arbroth (sic), Forfarshire, Scotland. Born Dec. 27, 1780. Died Oct. 17, 1871	Andrew Auld, a native of Linlithgow, Scotland. Born Sept. 8, 1799. Died Oct. 26, 1873

'Twa croanies (sic) frae the land of heather,
Are sleepin' here in death th'gether."

NAHIENAENA
(Sister of Kamehameha III.)
At the age of 10
From painting by R. Dampier,
Artist of H.M.S. *Blonde*.

In the harbour were three American merchant ships and several smaller vessels, two of which were native brigs mounted with 8 to 10 small guns. To the west of the harbour lay the wreck of a whale ship of 400 tons, from the port of London, wrecked two weeks ago during the night whilst waiting for the captain, who was on shore enjoying himself too freely. The crew saved themselves and her cargo of nearly 800 barrels of sperm oil, as also most of her spars, rigging, etc., have since been nearly all saved.

No one was allowed to go on shore today but the natives and his Lordship and the chaplain, who upon landing were met by crowds of the natives making the same howling noise as before when we were at the island of Mowee.

Poi and Raw Fish

Since we reached the islands, we have been much amused at the natives' simple manners, going to different parts of the ship, eagerly examining what they saw. At other times we were not a little disgusted when they sat down to their meals, eating raw fish with the gills and entrails and their fingers covered with blood. They appear to have no stated time for their meals, but ate when they felt inclined, sometimes as late as 12 o'clock at night, at other times by daylight in the morning. Their diet consisted of raw and dried fish and taro. The latter they generally pound with a mixture of cold water, on a thick board, with a stone, to nearly the same consistency as we do starch for linen. Then it is put into calabashes till it is wanted to be eaten. This they term poi, and is, when they can procure fish to eat with it, their constant and favorite food. When a family has its meals they sit on the ground in a circle, with their calabashes of poi in the centre. In turns they keep dipping the right hand forefinger, somewhat bent, into the same dish, and then thrust the finger into their mouths. The poi tasted to me sour and unpalatable, but when sweetened with sugar was greatly improved, and some on board liked it.

Native Mode of Measurement

The unusual size of our ship greatly attracted the natives, and when they came on board they began to fathom her length from stern to bow along the bulwark, at other times laying themselves flat on the quarter deck. Their knowledge of measurement appeared to us to be very imperfect, only amounting to a calculation of fathoms taken by the hands, dependent on memory. This is different from the Africans, who tally their numbers by notches on a stick.

At 8 o'clock p.m. we were again surprised to hear the gun from the fort, as is customary at that hour with the garrisons of other nations.

Kauikeaouli, Nahienaena and Kalanimoku

May 7. Went on shore with Lord Byron and most of the principal officers. On landing we were met by Capt. Charlton and a number of the chiefs ready to join us in a procession to Mr. Pitt (Kramaku's)[13] house for the purpose of being introduced. The chiefs were all dressed in European costume, and on our leaving the landing place, they in friendly manner placed one of us between every two of themselves, each having hold of our arm, till we reached Pitt's house, a small thatched hut pleasantly situated near the farther end of the town, within a large garden fenced in with high poles standing upright.

Pitt was dressed in black. His visage is thin and he is defective of an eye, but is by no means unpleasant in his manners. At the upper end of the hut was a platform raised about a foot from the floor, covered with the same kind of mats as the rest of the hut. Here the young king[14] and his sister,[15] whom we had brought from Mowee, were seated on a shabby sofa placed crossways, and having at the backs of it several handsome large feather plumes of various colours, customarily used by them in former times when at war and on high festivals.

Both looked delicate and of rather dark complexion. They had full nostrils and large mouths, but had fine open countenances, with good eyes and teeth, and not altogether wanting a sensibility of look that rendered them engaging. The king wore a short blue jacket with shirt and pantaloons, without shoes or stockings. His sister was in the said mourning dress from England.

From the king's sofa on the right, the principal chiefs and general officers stood in a line alongside the hut. From his sister on the left, lay on the mats by her sofa, Queen Kaumanna and two more of their father's[16] wives still living. After these three, stood, according to their seniority, five of their deceased brother's[17] wives, joined by a number of other female chiefesses, to the bottom of the hut. These were dressed in black canton crepe or silk of the same colour, with shoes and silk stockings and showy combs in their hair, everyway neatly dressed, contrary to what we all expected to find these people. When the ceremony of being introduced to the Regent (Billy Pitt) was over, through the

13. Kalaimoku or Kalanimoku, alias Billy Pitt.
14. Kauikeaouli (Kamehameha III).
15. Nahienaena.
16. Widows of Kamehameha I.
17. Widows of Kamehameha II.

KAUIKEAOULI
(Kamehameha III.), at the age of 11
From painting by R. Dampier,
Artist of H.M.S. *Blonde*.

Spaniard (Mr. Marin[18]) who acted as interpreter, the presents brought from England were then presented to His Majesty by Lord Byron.

Presents From England

The presents were a suit of full dress Windsor uniform, an elegant sword and a gold watch, with the arms of the king of Great Britain engraved on the back. Pitt was presented with a handsome gold ring. The king immediately tried on his clothes, and found it fitted as if made for him. He looked extremely well in them and was proud when he had on his cocked hat and sword, walking about speaking to some of his subjects.

When the presentation ceremony was over we were ordered to sit on chairs arranged for us in the middle of the hut. Country wine, made the year before, was offered to us by Mr. Marin. Some sat tasting the wine, and others got introduced to the females who were now sitting on mats. They could only answer with a smile, not knowing our language. The queens were all of enormous bulk, and by no means handsome.

The Horticultural Society of London

Being tired of waiting any longer at this mock formality, I slipped away unperceived, to look for plants, but being missed by Lord Byron, I was sent for to return, and was presented to the Regent as the person who had brought them plants from the Horticultural Society of London. He was informed that I wanted permission to collect the wild plants of the country for the Society. Mr. Pitt, Regent, kindly granted me full liberty to collect what plants I wished. Lord Byron, however, requested me not to begin collecting until the funeral was over.

Honolulu in 1825

In the afternoon, with some of the officers, I went for a walk through the town, which is situated on a sandy flat, with scarcely a tree for shade, except a few coco-nut trees in groups in places along the beach. The native huts are small and thatched with grass from top to bottom, but there are others inhab-

18. Francisco de Paula Marin, known to the Hawaiians as Manini. He is said to have had 52 children, thus carrying into effect King Kalakaua's motto "Hooulu lahui," "Increase the nation." Marin died 30th October, 1837, aged 64 years. Came to the islands about 1794.

HONOLULU
(Looking towards the Fort and Alewa Heights.)
Drawn by L. Choris.

ited by some Europeans and chiefs, which are covered over with mud, half a yard thick, to prevent accidents by fire, which so often occur in the others. These are better finished inside, and the floors spread over with mats made from rushes or the leaves of pandanus.

The huts of the poorer classes are mere hovels, having a low door placed where they creep on hands and knees to get out and in, with nothing to cover the opening but a piece of cloth or mat. Some of these into which we looked out of curiosity, had a hog or two tied up in a corner, and in others a dog nearly hairless from mange. The stench from having these animals live with them in the same hut was most offensive. There are as yet only four or five houses built after the European manner. One belongs to an American merchant, another to Mr. Pitt, not yet finished, part of which is at present converted into a guard house for a few naked soldiers who do duty at times by way of mounting guard in front of the king's hut, after a peculiar fashion of their own.

They do this in the following manner. Six or eight of these turn out together and form a line (not a straight one), and keep walking backwards and forwards, one after another, till they are dismissed. While they are in this way on duty, he that is last or foremost rings a small bell which is carried in the hand by way of signal for the others to turn either backwards or forwards. Their accoutrements are not all alike. Some have only a bayonet in their hand, held upright or reversed, just as suits their convenience, while another has an old rusty long barreled musket of American make, without a flint and sometimes a lock. Some have a cartouche box tied on behind with a piece of untanned goat skin, others have it in front and some have none at all.

The town of Hanarura contains about five or six hundred houses, and if the number of its inhabitants is taken at about ten to a hut, where they generally live together in families of two or three generations, they will amount to about 6000, which, I think, is underrated. Their huts are built without any regular form, enclosed with low mud walls, and a small garden, but without taro ponds the same as we saw at Mowee. Some of these gardens are cultivated with tobacco, Indian corn, water melons, pumpkins, etc., while others suffer weeds to grow and neglect cultivation, preferring to use them as a stockyard for hogs, goats, dogs and poultry.

The Harbour and Punchbowl Forts

Beside the harbour they have built a fort of mud and coral rock picked up at low tide. It is square and mounted with upward of 50 guns, many of which are 18 pounders, got from the Americans in exchange for sanders wood. From its situation it is not capable of long defense from an enemy at sea off the mouth of the harbour. It is at present in bad repair, and is in charge of an old Irishman, who has been on the island for many years.

About a mile above the town on the top of a hill[19] with the appearance of a volcano, they have another fort that mounts ten guns, which command the town and the taro ponds with other provisions cultivated in a large valley well watered by two rivers which run on each side till they meet in one behind the town.

Among the many Americans who live on the island, two keep public houses for the accommodation of strangers where they have managed to introduce a billiard table each, and are supplied with all kinds of spirits as well as wine at times from their country whale ships touching here for fresh provisions and water.

Sees the Hula

May 8. Sunday. Went on shore after dinner with several of the gentlemen from on board. Met Mr. Charlton, who agreed upon tomorrow for me to go with him to Mr. Pitt to arrange matters about my going to the woods collecting. Mr. Charlton offered me the use of three of his jackasses brought from England, to carry myself and specimens. These, I assured him, would be more hindrance than use, and asked him for three natives in preference, but in the end I was obliged to accept one jackass, as he would take no denial. He promised to have ready what natives I wanted as well as a guide.

In the cultivated grounds above the town, I noticed some good sugar canes, mostly of the purple striped kind, and a few patches of potatoes with weak stems not likely to be productive. Returned on board in the evening having been much amused observing the natives' simple manners and mode of dancing, which they accompany with a song and graceful motions of the arms and body, raising their voice at intervals to a high key, then again lowering it, without any given certain time that had in the least resemblance to music.

The inhabitants here all approached us without the least timidity like those we saw at the other islands, probably arising from greater intercourse with foreigners. Several could speak a good many words of English, to whom they are much attached, and often we heard the word "Britanee maitee,"

19. Puowaina or Punchbowl Hill.

which means "English very good" and superior to people of other nations, whom they look upon with jealousy.

Sandal Wood Trade

May 9.[20] Got nothing arranged as promised by Mr. Charlton as to my collecting, owing to the preparations making for the funeral. The natives load vessels annually with sanders[21] wood for which they get in exchange European and Chinese manufactured goods with every advantage to the foreigners. So eager, I was informed, have the traders been since the discovery of this native wood by one of themselves, named Brown,[22] in 1810, that every kind of scheme has been entered into for supplying the government with such articles as they thought would most attract notice and likely to be purchased. For this purpose they have introduced slight vessels built of fir, fitted with showy cabins with looking glasses, sofas with red morocco cushions, etc., and sold them at the enormous price of 80,000 dollars[23] to be paid for in sanders wood as it could be got from the different islands, without further trouble to themselves than the bringing it to the sea beach. For the traders themselves have small craft constantly in the harbour of Hanarura where they have on shore large store houses for the convenience of lading vessels. Besides vessels, the traders have furnished the natives with all sorts of naval and military stores, also small wooden frames of houses made in America, and light carts, some with two, others with four wheels. These scheming speculators have lately brought over a handsome carriage, hoping to barter it to advantage, but in this they have been disappointed, although the Regent and some of the queens were treated to a ride in it soon after it was landed. The natives still prefer the carts where they can lie down with more ease when they go to bathe. These carts are used but rarely on any other occasion, and are drawn by natives in preference to horses or mules of which they have several on which they often ride without saddles.

---

20. In the afternoon of May 9, 1825, a party of officers from the *Blonde* came ashore and marked out the first cricket ground that was probably ever used in the isles of the Pacific. In later years the Fijians, Tongans and Samoans became very fond of cricket, playing up to say 50 to 100 men on each side! In Hawaii cricket never caught on. It was too slow a game and baseball is preferred.

21. Sandalwood.

22. This Brown was not the Captain Brown who discovered Honolulu harbor and gave it the name of Fairhaven. The latter Brown was massacred by the natives in 1795.

23. e.g., *Cleopatra's Barge*, sold to Liholiho for $90,000, payable in sandalwood, and renamed *Haaheo o Hawaii—The Pride of Hawaii.* Two paintings of this vessel are exhibited in the Marine Room of the Peabody Museum of Salem, Mass.

Bathing the Chief Amusement

Bathing is their chief amusement and alone induces many of the higher ranks of them to leave their homes, where they spend most of their time sitting or lying down asleep on mats. But the whole tribe is so fond of bathing that the sea shore is seldom seen without numbers of both sexes swimming with perfect ease, as if some species of aquatic creatures.

Early American Settlers

Although the Americans have been on the islands for many years and enjoyed lucrative trade and made good fortunes, none of them have yet possessed lands of their own in this country, whether from the aversion of the chiefs to strangers or from the Americans caring only to trade in sanders wood and not take leases or grants of lands. Anyway they have but little improved the morals of the natives, and have neglected to teach them the arts of agriculture beyond what they already understood themselves when visited by Cook and Vancouver, viz., that of growing the taro root, their chief food. Even this was much neglected during the late king's reign, who, unlike his father (Tamahamaah) encouraged gambling and vices of all kinds. It is said of this late king that during his reign, till his departure for England, he indulged in every kind of debauchery and intemperance.

Liholiho and His Harem

The old Irishman who had charge of the guns at the forts, told me tales of him, and showed me the king's hut where he lived till he started for England. It was divided into two halves, one of these being again divided into separate apartments by low canvas screens for his six wives, who lived separately and never associated together or with the king, except one at a time. The said king had similar accommodations fitted up on a brig (the *Albatross*), where he sometimes went with his wives for a cruise to sea for two months together, taking the Irishman as his captain. During such a cruise he had hardly ever been free from intoxication.

The royal coffins were being cleared from the hold and were to be brought on deck the next day.

May 10. The coffins brought on deck, and those who wished, could look at the king and queen, for the outer case lids were removed. Arranged with

Lord Byron and Mr. Charlton to have a hut ashore, as I could do nothing with my plants on board for want of room. Learned that the funeral would take place early the following day.

Funeral of Liholiho and Kamamalu

May 11. The coffins were lowered over the ship's side into the launch. Those going on shore to the ceremony got into boats at 1 p.m., the boats forming a line ahead of the launch to tow her ashore. Lord Byron went in his gig on the right, the first lieutenant in the whale boat on the left, with colors in all the boats hoisted half mast. On leaving the ship 26 minute guns were fired, which was continued by the same number of guns from the forts on shore and a few from an American ship in harbour. The procession of boats towing the bodies had a grand and solemn appearance. On landing, the coffins were placed on two light four-wheeled carts, covered with native black tapa cloth, the head ends of the coffin in view, and drawn side by side by their late majesties' favorite domestics, preceded by our marines and the band playing slow marches. Next after the wagons came his present majesty, supported on the right by Lord Byron and on the left by Mr. Charlton, the British consul, all in full uniform.

After these came the lieutenant and gunroom messmates and the queens in black, but many of the latter without shoes and stockings. I followed with some passengers, and then the lower rank natives and at the end about 100 sailors, four deep, with a midshipman right and left. Thus we advanced between a line of native soldiers with rusty arms reversed, naked except for the maro, save a few that had on Russian military jackets and six on each side wore handsome coloured feather tippets. These twelve we concluded must be of higher rank. The five large feather plumes, used on the day of introduction, were carried before the band in a leaning posture, as they do when going into battle.

The procession halted at the missionaries' chapel, where our chaplain read a prayer in English, and then one of the missionaries did the same in the native language. Resuming our way, the procession reached Mr. Pitt's hut, between two and three o'clock. The marines formed a line in front outside the hut. The weighty coffins were taken from the wagons by the sailors and deposited on a platform inside the hut. Here the queens of the deceased king, joined by three of his father's still alive, and who are of higher rank, stood at the head of the coffins, and were joined by the wives and daughters of the chiefs forming a line on one side of the hut. Pitt, Boki[24] and other chiefs of consequence were admitted, together with the principal officers of the *Blonde* and a few Americans. No inferior class of natives were admitted nor had been allowed to join in the procession.

Boki's Autograph (from an original)

Then the chaplain gave out an anthem. It was accompanied by the band. A prayer followed, then the same performance by a missionary in the native tongue. Pitt, who had never joined the procession, remained at home in his arm chair, as usual dressed in black.

The whole ceremony was conducted with solemnity from the time of leaving the *Blonde* to the finish. One mistake, however, had been made. No invitation had been given to the few respectable Americans resident in the place; why is best known to those who had the conducting of the funeral, for they were left to form a small body by themselves on one side, away from the procession. It could not have been the wish of the chiefs, for they were seen often on their way to leave their places and join the Americans, especially Boki.

On the conclusion of the ceremony, all belonging to the ship were ordered back to their boats in marching order, with the band in front playing lively tunes. The natives broke through their soldiers' guard and clustered around us, evincing great astonishment at the great drum, and accompanying us to the harbour, where the ship's company were served with grog, and the officers had similar refreshment at Lord Byron's house. By five p.m. all were safely back on board. Mr. Dampier, the draftsman, had been stationed at an early hour where he could best make a drawing of the funeral procession,[25]

---

24. Boki, Boke or rather Poki, younger brother of Kalanimoku. He and his wife Liliha accompanied Liholiho to England. In *A Visit to the South Seas*, the Rev. C.S. Stewart mentions that when he revisited Honolulu in 1829, he found "the neat wooden building erected near the fort by the regent Kaahumanu, and occupied by Lord Byron during his visit, had been removed into the town, on the level ground, some distance from the water, near Mr. Jones; and, fitted with green blinds, a flagstaff and lookout, stands a conspicuous object, both from the water and on shore, as the Blonde Hotel, owned by Governor Boki."

25. Some of the drawings by Robert Dampier, artist on board the *Blonde*, appeared in the volume called *An Account of the Voyage of the* Blonde, which was compiled by a Mrs. Maria Graham. The same views and portraits of Hawaiian chiefs were also published in portfolio form.

KALANIMOKU
(Billy Pitt)
Drawn from life by L. Choris.

and is said to have already begun taking portraits of some of the queens with whom he has taken up his residence among them at their homes.

Botanizes in Nuuanu Valley

May 12. Had permission from Lord Byron to go on shore to collect; also the favour of a lad from the ship to assist. Took some salt provisions, biscuits and my bedding on shore at 3 p.m. Stored these at first in one of the rooms of a wooden house apportioned to Lord Byron during his stay, but afterwards I went to the hut of one of the natives (Manaware[26]), brought out with us from England, who had begged me that when he got home I would consider his house my home. On my applying for the fulfilment of his promise, he coolly forgot all about his obligations to me, and told me he knew of no hut for me, but that perhaps Pitt might find me one.

I then went on board Mr. Charlton's vessel[27] where he resided in preference to living on shore, but was again deeply disappointed, as he had Lord Byron and others at the time on board to dinner. When I met Mr. A. Bloxam,[28] who wanted to accompany me to the woods, and he proposed that we should both lodge in Lord Byron's house for the night.

May 13. Got up at 4 a.m., called the lad (Mantle) and by 5 we were ready for our journey, taking salt beef and biscuits with us and one of Mr. Charlton's asses to carry our packs. Mantle had charge of the ass. He could neither ride, drive nor lead the ass fast enough to keep up with us. It was therefore agreed to cast the ass adrift, leaving the pack saddle in one of the native huts. In crossing over the taro ponds to get to the woods, except being surrounded by low stone walls for several miles, I did not notice any difference in their cultivation from those I had seen at the island of Mowee. Here perhaps the ponds were rather better weeded. After having travelled for three miles we came to a fine, clear stream, where we halted to breakfast. We were surrounded by crowds of natives from the neighboring huts, who sat down on their hams, with us in the middle. They looked like so many starved dogs, staring at each mouthful of beef or biscuit that we took. Hoping to shame

26. Manuia.

27. The schooner *Active*, belonging to Capt. Richard Charlton, on board which he traded throughout the Pacific. It was this vessel that brought the Rev. Wm. Ellis to Honolulu, February 4th, 1823, on his second visit from Tahiti.

28. Andrew Bloxam acted as a sort of amateur naturalist on the *Blonde*, while his brother Rowland Bloxam was the ship's chaplain.

them away, we gave several of them food, but it only caused crowds to come till at last we had to tie up our provisions and move on.

During the time we stopped at the river, I gathered two specimens of polypodiums and three of convolvulus, besides a hibiscus with flowers, white inside and light purple out. We had not gone far, still followed by a number of natives, when we came on a chief digging up sweet potatoes. He addressed us in English and was very friendly, and on hearing of the object of our journey he gave Mr. Bloxam a young active boy to carry his traps till our return to town.

By 10 we began to enter the woods, having travelled about five miles most of the way through taro ponds and cultivated patches of sweet potatoes, bananas, water melons, etc., and in places near the river, a few low, unhealthy breadfruit trees, not in bearing.

Okolehao in 1825

The tea tree (*Dracaena*) of the country grew in all places, uncultivated in abundance. It was 2 to 4 feet high and about 1 1/2 inches in diameter. Beyond the provision grounds, where it grew in great abundance, were sheds where the chiefs, during the last king's reign, had been in the habit of distilling great quantities of spirits from the fermented liquor made from the roots, by means of large iron try-pots obtained from whale ships in exchange for provisions.[29]

Birds and Blossoms in Nuuanu Valley

On entering the woods we met with two trees of the *Eugenia malaccensis*, on which were a number of birds sucking its red blossoms. Mr. B. had the luck to shoot one of them, which he said was a species of humming bird.[30] My meeting with this kind of rose-apple, apparently growing here indigenous,

29. The art of distilling spirituous liquors from ti (*Dracaena*) root is said to have been introduced into Hawaii before 1800. Under the 18th Amendment to the U.S. Constitution, the manufacture of all kinds of spirituous or malt liquors is made illegal in the United States, of which Hawaii now forms a part. Notwithstanding this, the illicit distillation of spirits made from rice, pineapples, maize, etc., is carried on to a much greater extent than in former years, and during the year 1921 over $60,000 in fines were collected in Hawaii from transgressors of the prohibition law. The distillation of spirits from ti root was introduced into Tahiti in 1798 by two Sandwich Islanders who had deserted from the British N.W. Coast fur trader *Nautilus*.

30. There are no humming birds in the Hawaiian Islands.

so far from habitations, rather surprised me in the Sandwich Islands. The other trees seen were common,—*Aleurites*, and a species of *Acacia*, used by the natives for making their canoes and paddles. On the ground below were mixed a variety of handsome ferns in all kind of places, moist and dry. Further towards the centre of the island, the trees became more lofty and the ground below them more shady and damp, where there appeared several species of *Psychotrias* and *Beselerias* with two or three tall-growing *Lobelias* with splendid clusters of flowers. There were also three kinds of *Metrosideros* with rich bunches of scarlet flowers. These were covered with birds, sucking honey from the blossoms, which we shot, but could not afterwards find, owing to the thick growth of ferns, plants, etc.

Nuuanu Pali

We gained the head of the valley at 5 p.m., having travelled about ten miles, when we had a splendid view of the Pacific Ocean on both sides of the island. Here we had a cool breeze, occasioned by the valley terminating in a narrow gap, overhung on each side by high abrupt cliffs covered with tufts of *Lycopodium cernum*, so common within the tropics, and two or three kinds of *Vacciniums*, besides other low shrubby plants which clothe this mountain, at least 3000 feet to its summit, with pleasant verdure. At this point there is a path which leads down to the valley on the other side of the island.

While we were at this place, which is a thoroughfare for the natives crossing the island, we were surprised to be overtaken by a man, better dressed than the rest of the natives, who accosted us in English and very familiarly entered into conversation. He told us he was a native of Otahiti, which he had left when a boy to serve on a whale ship. Afterwards he was in the British navy, till he was wounded at the battle of Algiers, when he was discharged as unfit for service with a pension of twenty-five pounds a year. Having learned from this man that we could travel by the sea coast round the east side of the island to the town next day, we decided to remain in some hut for the night. But in order to advance farther, we had to make the most difficult descent to the bottom of the precipice by a winding rocky path, in places for several yards together quite perpendicular. We had to take off our shoes and to scramble as best we could, at times backwards on our hands and knees. The natives in our company carried their loads seemingly with perfect ease, and enjoyed heartily seeing us in the least terror to go on, coming to our assistance to prevent our falling. Fortunately we got to the bottom without any accident, and soon came to two or three small huts where we obtained lodgings for the night in the hut of a Bengal black, who had been wrecked in an Indiaman on the coast some years ago, and had since then lived on the island with a native woman, by whom he had several children. He spoke good English, and told us he was a tailor, and sometimes acted as cook for the king before he went to England.

Visits Kaneohe

As it was not yet dark, we proposed a walk to the sea side, about 1 1/2 miles distant, while our host cooked us supper. On our way we passed several native huts with little patches of cultivated ground, chiefly planted with Brousonettia, from the bark of which they make the tapa cloths. They also cultivate a plant which they name None,[31] for the sake of its fruit, which yields their favorite yellow dye for the tapa cloths. In the hollows were some taro ponds and several groves of healthy trees of the rose apple, growing apparently without any cultivation and coming into fruit. The bay is open and exposed and full of rocks in many places above water, which renders it unsafe for vessels to anchor. It is full of fish.

The high ridge of mountains which runs nearly in the centre of the island, terminates on this side in high abrupt cliffs. Along the coast for four or five miles back into the interior, the country is, as in the neighborhood of Hanarura, left everywhere uncultivated as nature had formed it, excepting the small patches round the native huts. Although there is good pasture, none of the government cattle have yet been sent to this side from Hanarura to feed.

On our return we found some taro root ready baked under ground with hot stones, and a small fresh water fish of the mullet kind cooked in the same way, enclosed in the leaf of a tea tree. We slept on mats by the side of the Bengal black and his wife, with her father and mother and the rest of the family, in all, upwards of a dozen men, women and children, besides several dogs, thickly stowed together at one end of the hut without distinction.

May 14. Got up at 4 a.m., after a restless night from fleas and cold, the hut being so open that light could be seen through it. Informed by our landlord that going round the island, as we intended and were told by the Otahitian, would take but one day, would, in reality, take two days. For want of provisions, we were reluctantly obliged to return by the road we had come. We left our host at 5 a.m., the dew still on the grass, after giving him a trifle for his last night's trouble.

---

31. Noni (*Morinda citrifolia*), a plant found throughout Polynesia, Malaysia, etc.

Climbs the Nuuanu Pali

We overtook a number of natives at the precipice, which we had found so difficult to descend. Most of the natives were loaded with provisions for the chiefs, such as large hogs, tied together by the four feet and carried on the back, at other times led or driven. Others with bundles of taro or large calabashes of poi, secured to each end of a staff thrown across the shoulder, and travelling thus with a short shuffling step, stopping to rest themselves every now and then, and to have a draw in turn from a wooden tobacco pipe, and then resume their journey as before.

Tatooing and Hair Dressing

They have no calculation of time, beyond the rising and setting of the sun, and only care to gain some hut for the night, being too superstitious to travel after dark. Both sexes are generally tatooed irregularly, with figures of goats, muskets and even letters of the alphabet. Name and birthplace with date of the year are often seen tatooed along the arm. Many of the men shave their heads, and cut their hair in the form of a helmet, the crest of which is often stained with lime, so as to be of a light whitish colour. The women esteem it cut short, with a rim over the forehead bleached white and standing up in front like bristles. Sometimes a long curl is preserved in the middle of the forehead, which is combed backwards. Some suffer their hair to grow and tie it up behind in a bunch. Many of the females, different from any of the other sex, have a tatooed line about two inches broad inside the thigh down as low as the ankle, where it terminates in the form of a ring generally on the right foot. Most of them had a looking glass and a wooden tobacco pipe tied round the neck in a handkerchief or piece of tapa cloth. They are very fond of smoking, and are seldom seen without a pipe, and curiously enough the habit among them is, though most of both sexes have and carry a pipe, for one of them to light his or her pipe, and after a few draws by the person who filled it, to pass it along without distinction of persons, until all have had a draw at it.

By ten o'clock we were nearly through the wood, when Mr. Bloxam, impatient of waiting for us, parted from us, he to return home before the sun's heat increased, while I remained behind in order to collect more plants not met with yesterday. When I arrived in Hanarura, I found that no place had been arranged for me, while I am out in the woods. Got all my traps moved on board and lay out my plant specimens on paper.

May 15. Sunday. Church service on board at 10. Spent most of the day

TATOOED HAWAIIAN CHIEF
Playing the game of balancing on stone ball.

laying out specimens, sorting seed, etc. The surveyor, purser and midshipman, in the whale boat, were upset in the surf going on shore, and had a twenty minutes ducking, till another boat went to their assistance.

May 16. Went on shore with Mr. Forder, who intends to remain ashore and draw plants, and take care of the hut provided for me, while I am out in the woods. Got all my traps moved into my hut, which is about half a mile east of the town, and has never been tenanted before. Shifted my specimens, while Mantle slung our hammocks across the corners of the hut. Before dusk, Lord Byron and the surgeon rode out to ask if I would care to go with them and a party to Pearl River by water, and if so, had better breakfast with them at 8 a.m.

Trip to Pearl River or Harbour

May 17. Joined Lord Byron's party, with Mantle carrying my traps. We did not embark until noon. After two hours sailing along the coast, we entered the mouth of the Pearl River, which divides itself into several branches, forming two islands. One which is smaller than the other is called Rabbit Island, from a person, the name of Marin, a Spaniard, residing at Hanarura, having put rabbits on it some years ago. The rabbits have since increased in numbers.

It became so calm, that his Lordship, Mr. Charlton, and the Bloxams left us in the launch, and rowed in the small boat in tow, and soon disappeared from sight. We waited in suspense, hour after hour, not knowing the several branches of the river, nor where we were to spend the night. The boat party pulling into one branch of the river, the other in which I was tacking about from bank to bank till the boaters hauled their boat ashore and we cast anchor. Both parties were opposite each other on Rabbit Island, but ignorant of the fact, till on walking about the island, the parties met. One hut was noticed, and those on the island made for it, but the launch having the ladies and some others on board, got up anchor and sailed round to the hut, where with the help of canoes, they all landed. The ladies were somewhat discontented, but after a good dinner partaken sitting on mats spread on the grass, harmony was restored.

At dusk we embarked to cross to a larger hut. Landed at 8 p.m. At ten o'clock two old men entered our hut to play the hura dance on a couple of bottle shaped gourds. They took a sitting posture, beating time on the gourds with the palms of their hands, accompanied by a song made up about the late king.

About 11, we all retired to rest, lying down beside each other on mats, some with pumpkins or what else they could get for a pillow. The ladies got themselves screened off in a corner with a flag without any other accommodation.

Pearl River is about seven miles west of Hanarura, and is improperly called a river, being rather inlets from the sea, branching off in different directions. There are three chief branches, named by the surveyors, the East, Middle and West Lochs. The entrance to Pearl River is very narrow and shallow, and in its present state it is fit for very small vessels to enter, but over the bar there is deep water, and in the channel leading to the lochs there are from 7 to 20 fathoms. The lochs themselves are rather shallow.

The coast from Hanarura to the west of Pearl River possesses no variety of plants beyond two or three species, such as Argemones, Portulacas, and a few other little annuals, intermixed with the common long grass so plentiful everywhere on the coast round the island.

Oysters

The oysters that are found in Pearl River are small and insipid and of no value or consequence.

Returns By Land

May 18. Got up at 4 a.m., after a restless night, having been tormented with fleas. Departed with my man Mantle, leaving the rest yet asleep. But after travelling about three miles, the path which we had first struck terminated, and the grass became longer and more difficult to travel over. At last, after another three miles, we got so entangled with creeping plants running a little above the ground beneath the grass, that Mantle, who was stockingless, shed tears, complaining of his ankles, and refused to go on. Being yet five miles from the woods, and not having sufficient provisions for two days, we were forced to return to the town by a path leading through taro ponds, some distance inland from the coast.

On the path we had left near the Pearl River, we saw several thickly inhabited huts, situated on the side of a ravine stocked with bananas, taro and healthy breadfruit trees just forming their fruit. Here we met with an old Englishman, who told us there was on the opposite side of the ravine a large river coming out under the ground. We went to the place and found that what he had told us was correct, and stood admiring the subterranean stream of fine, cool water. Its source was rapid, forming a cascade nearly 20 feet in height, having ferns and mosses on its sides. In the grounds of the natives, I saw plenty of the awa plant (*Piper*) mentioned in the history of these islands, as being destructive to the health of the natives when used to excess, owing to its intoxicating qualities. I obtained several specimens of it in flower.

The old man informed me that he had been on the island over sixteen years, and that the grounds we were then upon, belonged to Boki, and had been in his charge for ten years. Upon Boki going to England with the king, another chief had turned him away, and taken all his little ground from him, so that he had been forced to live on the charity of the natives.

Ewa District

The neighbourhood of the Pearl River is very extensive, rising backwards with a gentle slope towards the woods, but is without cultivation, except round the outskirts to about half a mile from the water. The country is divided into separate farms or allotments belonging to the chiefs, and enclosed with walls from four to six feet high, made of a mixture of mud and stone. The poorer natives live on these farms, also a few ragged foreigners who have a hut with a small spot of ground given them, for which they must work for the chiefs a certain number of days besides paying an annual rent in dogs, hogs, goats, poultry and tapa cloths, which they have to carry to whatever spot their master is then living on the island. On the least neglect to perform these demands, they are turned away and deprived of whatever stock, etc., they may possess. Such is the present despotic or absolute law in the Sandwich Islands. This is corroborated by all foreigners met with at different times, who, on our arrival, hoped that Lord Byron would render them their little property more secure in future. Unfortunately they must wait till the British Consul helps them, as we have no authority to interfere with the laws of the country.

On our way home we noticed that the country on the side towards the woods still remained uncultivated, also towards the sea coast, except the lower ends of the small valleys which are cultivated with the taro in ponds, which much resemble peat mosses that had been worked and afterwards allowed to get full of stagnant water. There is no convenient road to travel anywhere on the island. We met with another subterranean river at the side of one of the hollows, larger than the other, but of no great fall after its appearance from underground.

Moanalua Hill

By 4 p.m. we gained the summit of a high hill, thickly covered with tufts of long grass. It lies within three miles of Hanarura. There is a burying ground of the natives at the top, which was formerly where the chiefs of high rank had a morai.[32] At the bottom towards the sea, there is a circular salt pond,[33] nearly two miles in circumference, surrounded by low conical hills. In places on the sides of a valley leading to the pond from the interior, are several huts of the natives with taro ponds and a large grove of coco-nut trees, apparently very old from their great height and mossy appearance. We reached town about six o'clock having travelled twenty miles since morning without much success, being too near the coast to meet with a variety of plants. We learnt, however, a good deal about the present mode of life of the natives, and the manner in which they continue to cultivate their grounds, differing but little, if any, from the descriptions given by Capt. Cook and others.

May 19. Fine. Saw to my specimens. Lord Byron and the surgeon called to hear particulars of my journey home from Pearl River. American missionaries called and invited me to return the visit. Mantle still complained of his feet, but will go in the morning to the woods with me. Mr. Bloxam accompanied me to the woods, but by 8 a.m. he said he had shot enough birds to skin and would go home. I asked him to shoot a few for me, as he had enough for himself, but he refused, saying all his duplicates were for Lord Byron.

Climbs "Tantalus" Mountain

As we advanced to the wood, we met with a multiplicity of ferns, many of them different from those I had seen on the 13th and 14th in Hanarura valley. By 12 o'clock we had gained the summit of the highest hill fronting the bay, and it rained in torrents. Here I took the temperature of the air, which stood at 69. This hill is over 2000 feet above sea level, but there are others beyond it in the centre of the island of much greater height. Two miles beyond this high hill, the wood became difficult to penetrate, and being so wet, we set out homewards. The ferns of different kinds met with during the day exceed twelve. We met also with a considerable variety of plants, among which were a parasitical species of *Metrosideros* with large scarlet flowers and two or three lobelias with handsome flowers.

Land Shells

I met also for the first time with small land shells, having a variety of rich striped colours. These shells were found chiefly on the leaves of the wild

32. In Hawaiian "heiau."

33. Known as Aliapaakai.

tapa (*Brousonettia*) of which there are several species with different coloured flowers. At the huts near the wood, we saw some natives eating raw sweet potatoes, just taken up from the ground and not even washed.

Got home at six o'clock. Lord Byron called to say he was riding farther on towards Diamond Hill, and would I meet him the next day at Pitt's to see the plants, which had been brought out from England, properly transplanted.

Explores Diamond Head or Hill

May 21. Met Lord Byron, according to promise, at Mr. Pitt's to transplant the fruit trees brought from England, but when I got there found he had not come. After waiting some time I went to his house, and saw Mr. Charlton, who doubted if anything could be done then, as Mr. Pitt and the other chiefs were then asleep, and would not get up on any account whatever. I was advised not to wait on the chance of finishing the planting, but to go, as I had intended, to Diamond Hill. So Mantle and I set off and were joined by a native who offered himself as a guide. By 6:30 we had gained the summit, which is high and steep, without anything growing on it but tufts of dry grass in loose sand, which came up very easily and rendered the ascent more difficult. Diamond Hill forms a headland near the sea, about three miles from Hanarura, and has at some former period been an old volcanic crater, now extinct. In the centre is a level flat, two acres in size, covered with longer grass than the external declivity. On the inside part next to the sea, the depth is upwards of 500 feet, counting from the narrow ridge round the top, which is almost circular. Some parts of the rocks had the appearance of common quartz; others resembled burnt limestone, turned soft and white.

There are many natives' huts along the coast on both sides of the hill, surrounded by banana and coco-nut trees. Coco-nut trees seldom grow far from the coast. Near Hanarura is a similar crater, now a fort, with several guns placed along the ridge. A salute was fired from this fort when we entered the harbour.

On our way home, in a high wind, my hat was blown off, and Mantle went down the cliff to retrieve it. It was nearly dark and he called out to myself and the native guide not to leave him behind. Afterwards he told me that in picking up the hat, he had picked up a large human skull. At first he had not noticed what he had picked up till he saw the hollow part of the eyes, and had called out in fright and thrown the skull away. On my expostulating he said he could get plenty more, as there were a lot there. This was explained to me by the American I had hired as cook and who had been on the island for six years, by his telling me that in former times, the chiefs used to take their criminals upon the top of Diamond Hill to put them to death by throwing them over the precipice, where they were left unburied in the hollow, and thus Mantle had met with many skulls.

Tamaahamahaa Alias "The Duke of Wellington"

May 23. At 4 p.m., the Governor of Attooi called on me. He was busy quelling the insurrection on that island about two months before our arrival. He is also styled the Duke of Wellington, the conqueror of Attooi. He is a young man of about 25, stout and good looking, in comparison to the rest of his countrymen. He was brought up by Boki, who was in England, and formerly, until made a governor and general, went under the name of Tamaahamahaa. I placed before him a nearly full gin case bottle of rum and a tumbler, which he finished by himself in less than two hours time, and went home but very little intoxicated. I took care, however, never to make him quite as free with my bottle again.

Visits Manoa Valley

May 24. Went to the woods through a valley near Diamond Hill, named by me Wilkinson's,[34] from its being the place where the unfortunate man lived that we had brought out from England with us in the suite of the Sandwich Islanders in the prospect of improving the state of agriculture here. But he found that he had to do with ignorant, unfeeling masters, who instead of employing him to improve their farms, with a liberal salary, had only given him some waste land in the valley to cultivate for his own, during their own pleasure. On passing through his land I found he was ill in a tent, having no hut built yet. He lay on a mattress on the ground, suffering from diarrhoea, and reduced to a skeleton, and looking the picture of death. I suggested that he should apply to Lord Byron for a passage home with us, but he said he would rather die where he was than return to England. His land was near the head of the valley in a kind of amphitheatre overhung by volcanic ridges, thickly covered with candle nut (*aleurites*) and other trees, that occasion much rain at night.

34. Wilkinson is credited with having been the first person to cultivate coffee in the Hawaiian group. The first plantation was started at the head of Manoa valley, Oahu, with seed or young plants obtained at St. Catherines, Brazil, where H.M.S. *Blonde* touched on the way out from England. Died March, 1827, before reaping the reward for his labor.

In several parts of this valley, I met with great quantities of turmeric, ginger, awa, and one specimen of canna, growing wild among the grass. The ginger is met with in abundance everywhere on coming to within half a mile of the woods, but afterwards it disappears on getting some miles further on. The roots are generally bitter, and have not much of the taste of cultivated ginger. When in the wood, I met with four more different species of *Metrosideros*, and several other plants which I had not seen before. At the head of the valley at the side of a small river which descends from the mountains, are trees of the *Eugenia malaccensis*, forming a wood by themselves. On my return home I saw hundreds of the natives preparing the ground near where I lived, for Mr. Pitt, in order to be planted with sweet potatoes. They did this by pulling up the grass with their hands, and others using a short stick to loosen the ground. They have as yet no other description of tools.

Two of Kamehameha's Widows Call on Macrae

May 25. Two of the late Tamahamaah's queens called on their way to bathe. Both of their majesties on coming to the door in a cart drawn by several natives, were tilted out without ceremony, like so much rubbish, being unable to get out in any other way, owing to their enormous size. They were dressed in Canton crapes, made in European fashion, with coarse straw bonnets and old slippers, but no stockings.

Mr. Forder to amuse them, showed the drawings of plants, etc., that he had made, and they asked to be given them all, and when refused, they asked for something else. They are very covetous. When they went, they got into the cart, laid hold of the front part standing up together behind, and in this manner the box (cart) was raised up with them face downwards, and their legs, nearly the size of a man's body, hanging over the hind part of the cart, which was too short for their length.

Looks For Sandal Wood

May 26. Showers of rain. Went to the valley above Diamond Hill looking for sanders wood. It is this kind of wood which has bestowed so much wealth on its owner, and caused the neglect of cultivation in the islands since its discovery, owing to the natives being taken away from their homes for weeks together to procure it.

In the wood at the head of this valley, I met with many plants not seen before, particularly ferns and *Metrosideros*. The latter formed the greatest number of the largest trees I had seen, except the *Acacia* from which the native canoes are made. The sanders wood I met with grew only to a middling sized tree, and was too early in the season for flowering. In the forest, bananas grow wild to a great size, with many large bunches of fruit on them, which the natives with me cut and roasted. One which I tasted was bitter, but by no means unpalatable. The fruit of the wild banana is the chief food of the natives when they are in the woods collecting sanders wood or for the purpose of rough-hewing their canoes light enough to enable them to drag them to the sea shore. This latter operation often takes several days.

Native Method of Obtaining Fire

Today for the first time I saw the natives light a fire to roast their bananas, by rubbing two dried sticks upon each other, that catch fire by friction, after rubbing them for about a minute.

May 27. Fine, but showery. Hearing that the ship would sail in a few days for Owyhee, I went to Lord Byron to ascertain if he would return to this island again before he left the islands for Otahite, for if so, I must have all my plant boxes on board. He said he could only tell he was leaving in a few days.

John Young and Isaac Davis

He introduced me to Mr. Young. His name along with that of his companion, Mr. Davis, have long been distinguished in the history of these islands. Mr. Young had just come from Owyhee. He is now fast sinking under the infirmities of age. Mr. Davis is dead and buried in the Europeans' burying place under a grove of coco-nut trees near where I live, and on his simple monument is engraved

"The remains of Mr. Isaac Davis,
who died at this island
April 1810, aged 52 years."

On his decease, he had possessed many lands or farms in the different islands, given him by Tamahamaah for his faithful services, which afterwards became the property of his wife and daughters, but some have lately been taken from them by some of the covetous chiefs now in power.

JOHN YOUNG
("Olohana"—"All hands.")
The friend and Companion-in-Arms of Kamehameha.
Born Lancashire, England, 1742; died Honolulu, December 17, 1835.
Aged 93 years.

Magic Lantern Show Banned By Missionaries

Lord Byron had a magic lantern show at the king's hut, but owing to the religious fanaticism of the American Methodists, the king was prevented from being present.[35] These missionaries, many of them being but illiterate mechanics, possess what power they please over the credulity of the natives, and have already carried their system of religion too far to be upheld.

Plants Brought From England

May 28. Met Mr. Charlton at 6 a.m. at Mr. Pitt's hut to see, transplanted, the plants that were brought out from England. I found the natives had already finished planting out the tropical plants, among which I had several *Orchidacae* from St. Catharines, doing very well in their box. Although my *Orchidacae* intended for Mr. Lindley[36] were now partly destroyed, my feelings at the moment were nothing compared to what they were when I saw the natives tear up my other plants by the stems, as a gardener would cabbage stumps for the dung heap. Mr. Charlton was not present, so I requested Mr. Pitt to stop having anything done until I returned with the Consul, at which Mr. Pitt only laughed at my eagerness.

Marin (Manini) the Spaniard

When I returned, I found that self-conceited Spaniard, Marin, helping to pull the plants by the stems, and continued doing so in my presence. I remonstrated, but Mr. Charlton drew me to one side and said "I had better leave the fellow to himself to do with the plants as he thought proper," to which I agreed, on condition that Mr. Charlton would explain to strangers touching here that I had no hand in the transplanting. I then took up my St. Catherines Orchidacae and replanted them in the box from which they had been taken, and then proceeded to make a list of what plants had been introduced

35. In regard to this magic lantern incident, the missionaries' side of the question may be learned by consulting the following works, viz., *An Examination of Charges against the American Missionaries at the Sandwich Islands as alleged in the Voyage of the Ship* Blonde *and in the* London Quarterly Review, Cambridge, Mass., 1827; also *A Defence of the Missions in the South Sea and Sandwich Islands, etc.*, by William Orme. London, 1827.

36. John Lindley, the well known English botanist and horticulturist, who during his lifetime was the mainspring of the Horticultural Society of London. It was under the auspices of this society that Macrae had been sent out in charge of a variety of plants to be distributed at the Sandwich Islands.

alive, but found the natives had stolen all the leaden numbers attached to each plant, as well as the iron hooks that kept the lids of the boxes open when the plants required air on the passage from England.

Alexander Adams

May 29. Sunday. Fine. Went with our purser to the house of the harbour pilot (Mr. Adams) to accompany them to the other side of the island, to see the steep cliffs where I and Mr. A.B. had difficulty in passing on the 13th. Mr. Adams amused us throughout the day relating curious anecdotes of the late Tamahamaah and his queens, particularly about Kaumanna. Adams had been on the islands since 1809,[37] sometimes conveying sanders wood to China, at other times to the North-West coasts of America with cargoes of salt. He brought home a couple of deer[38] the last time with a view of their thriving in

37. It was owing to the intercession of three British seamen, viz., Alexander Adams, John Young and Isaac Davis, who had been trusted councillors of Kamehameha I, and who had been continued in the same capacity by his son Liholiho, that the first contingent of American missionaries, after being confined on board the *Thaddeus* at Kailua in 1820 for two weeks, were finally allowed by Liholiho and his council of chiefs to land and settle on the islands. In the *Sandwich Islands Magazine* for 1856, page 78, Judge Abraham Fornander, the leading Hawaiian historian, gives his opinion on this point as under, viz., "Though all these probably had an influence more or less determining the decision of the king and queen-regent, yet after all the most direct influence that decided this admission of the missionaries and their permanent residence, was the intercession of a few Englishmen, long resident on the islands, and without whose names any account of the first period of Hawaiian civilization would be imperfect and unintelligible. That their names have not yet had honorable mention in this connection by those who have assumed to write on Hawaiian history, is one of the sins of omission which we would charitably overlook if we could. Those men were Young, Davis and Adams. The two first were settled here before, and contributed in no small degree to the conquest and its results; the latter is still living to reap that tribute of respect, that tardy justice, which those who have neglected to render, whom he and the others assisted in their hour of need and cheered in their moments of doubt."
38. Deer were not the only kind of wild animals introduced into Honolulu during the reign of Liholiho. In 1822, that monarch dispatched his American-built brig, the *Sunbeam*, commanded by an Englishman, Captain John Bowles, and manned by Sandwich Islanders, to St. Peter and St. Pauls (Petropaulovski), Kamtschatka, with a cargo of salt as a present to his imperial brother, the Czar of Russia. In return for this gift on the part of the Sandwich Islands king, the governor of St. Peter and St. Paul's, who was then Captain Ricord, an Englishman, gave such articles as seemed most desirable, including some animals, with a view of propagating the breed. Amongst the animals were two Siberian bears, but what became of them when landed at Honolulu is now forgotten. Some details about this voyage of the *Sunbeam* in search of bears will be found in Captain John Dundas Cochrane, R.N.'s, *Narrative of a Pedestrian Journey, etc.* 2nd ed. London 1824, pp. 425–7, and an amusing sketch called "Kamehameha II's Bears," by a Russian author, V. Poliakoff, appeared in Chambers' *Journal* for 1st March, 1920.

CAPTAIN ALEXANDER ADAMS,
Designer of the Hawaiian Flag.
Born at Arbroath, Scotland, December 27, 1780.
Died at Honolulu, October 27, 1871.
Aged 91 years.

the islands, but they had not long been suffered to go at large in Hanarura valley, when Pitt happening to be unwell, fancied that the flesh of the deer would do him good, and one of them was killed for him to taste. This he found so much to his liking that he ordered the other one to be killed, thus ending the life of poor Adams' deer.

Battle of Nuuanu Valley

On our way through the valley of Hanarura or Nuuanu, as it is often called, we were shown the spot where the king of Woahoo's (Tereaboo) head general was killed after the invasion of Tamahamaah from Owhyee to conquer the island. It happened as follows: When Tamahamaah with Young and Davis and the rest of his army had landed from their small fleet in the harbour, without opposition from Tereaboo, they found that the latter had collected his forces above the town in Hanarura valley. Tamahamaah could not have wished for a better situation or one more favourable to his purpose, the valley being overhung by ridges on each side, which were left unoccupied by the enemy. Tamahamaah, without any opposition from the enemy, placed a number of his men on the side ridges, and then he himself, accompanied by Young, Davis and the greater part of his army, took up their position in the center of the valley. They had with them only one small swivel and a few firearms, the rest being armed with spears and clubs. Mr. Davis, who had the swivel, somewhat singular to relate, killed Tereaboo's general on his firing the first shot, before the engagement had scarcely begun. When this happened, as is usual with these natives, they instantly got into confusion and retreated. Tamahamaah pursued them up the valley, and on coming to the precipice they threw themselves over and were found in the thousands, lifeless at the bottom of the cliff.

Thus did Tamahamaah, with the help of Young and Davis, and with hardly any firearms, easily conquer this important island, which may now be considered the first of the Sandwich Islands on account of its good harbour. The king of Woahoo fled to the mountains, being convinced that the custom of putting the vanquished to death would be practiced upon him. "I must die," he said to one of his friends, "for I will not let Tamahamaah enjoy this triumph. I will sacrifice myself to the gods." His corpse was afterwards found in a cave in the mountains.[39]

---

39. This account of the death of King Kalanikupule differs from that given by Fornander in

I managed to shoot some birds, one being particularly handsome. Its feathers were all red,[40] and it is only met with when sucking the red blossoms of the *Metrosideros*.

May 30. Fine, with light showers of rain. Therm. 6 a.m., 76; 12 noon, 86; 7 p.m., 76.

Alexander Adams Helps Macrae

May 31. Went with Mr. Charlton and Mr. Pitt to get my empty plant boxes taken to a carpenter to be repaired. Pitt promised natives to help, but none came, so in despair, I went to Mr. Charlton to go with me to procure assistance for which I was willing to pay rather than waste more time. However we met Mr. Adams, the pilot, who kindly borrowed a cart and got his own people and came with them and we soon got the whole lot of boxes removed.

The plants brought from England had remained without any shade from the sun. Through the neglect of these people, I found only one grape vine with any leaves on it. Plants which at the expense of the Horticultural Society, had been brought 15,000 miles by sea through various climates: Marin, the Spaniard, previously mentioned, has had a grape vine since 1814, which he brought from California, where he had been with Adams as linguist, and had planted in his own ground near Hanarura,[41] and got the king to perform the charm of taboo to prevent any being stolen. This taboo still continues in force to the present time, but although he has cultivated the vine with success for the last two years, and made more than two casks of wine from the fruit, sooner than give the slips to his neighbors, he has burnt the prunings every year.

But Mr. Charlton has given to different individuals, several fine grape vines which he brought with him from the port of Valparaiso, and also an assortment of vegetable seeds, the gift of the Horticultural Society, London. Mr.

---

the *Polynesian Race Vol. II*, p. 348. There it is stated that the king was captured in the mountains of Waipio, Oahu, brought to Kamehameha and sacrificed to the latter's god Kukailimoku. Probably the truth is that Kalanikupule's corpse was found in a cave by those hunting for him, and brought to Honolulu, where it would be offered up as an offering to Kamehameha's god.

40. The Iiwi (*Vestiaria coccinea*) or the apapane (*Himatione sanguinea*). Native birds are getting very scarce on Oahu and the other islands of the group. About 1910, or so, the writer noticed on two or three occasions, an iiwi in his garden situated on Prospect Street, on the south side of Punchbowl Hill. This is about one-third of a mile from the business centre of Honolulu.

41. At Pauoa Valley.

Charlton intends soon to reside with his family at Woahoo, where he intends to grow coffee and cotton, which have so far been neglected. In such a favourable climate, they, as well as cocoa and sugar, may be brought to perfection.

Queen Kaahumanu

June 1. Returning from the town, I saw Queen Kaumana in her four-wheeled cart being dragged to the top of a small hill by natives. The cart was afterwards pushed off at the top and allowed to roll down hill by itself, with her in it. This ludicrous sort of amusement was always accompanied with much shouting on the part of the natives.

June 2. Queen Kaumanna and Pio,[42] with several female attendants, called in their carts. They were very inquisitive to know if Mr. Forder and I were married men. Being told, yes, they wanted to know the number in our families. Mr. Forder, who kept up the conversation, said he had six and I none, on which they said I could only have one wife. After a few more questions, they left us to bathe, as usual, in the fish pond, near here.

A little before dusk, the king called on horseback, accompanied by a little boy brought out from England by Boki. His majesty was dressed in a short blue jacket and pantaloons to match, but without hat, waistcoat, shoes or stockings. Afterwards we saw his sister, the young princess, carried home from bathing on a native boy's back, some distance in front of the old queens, who were dragged in the carts.

June 3. The ship sails for Owhyee on the 7th, so I am instructed to have my traps on board again this evening. Packed up my specimens and got everything on board by dusk.

June 4. Got my things stowed away amidst the usual noise of a man-o'-war. At 5 p.m., one of the gentlemen hooked a shark alongside which measured 11 feet in length and five in circumference, and when cut open, in its stomach were found a large hook and chain carried away the day before, a bullock's foot, part of a pig's head, and many bones which had been thrown over the ship's side, where he had been noticed for several days past. The Sandwich people who were going with us to Owhyee sent on board a quantity of luggage and provisions for their passage.

June 5. Sunday. Fine. Church service on board as usual at 10 a.m.

June 6. Went on shore. In the town saw a large pig roasted with hot

42. Piia, alias Lydia Namahana, sister of Kaahumanu.

stones in the way customary with the South Sea Islanders, and as mentioned by Capt. Cook and others.

Women Suckle Pigs and Dogs

Shortly afterwards, I noticed a young woman walking along the street, and at the same time suckling several puppies that were wrapped up in a piece of tapa cloth hanging round her shoulders and breasts. The custom of suckling dogs and pigs is common to the natives of the Sandwich Islands. These animals are held by them in great estimation, little inferior to their own offspring, and my journeys to the woods in search of plants often afforded me an opportunity of being an eye witness to this habit. I often saw them feeding the young pigs and dogs with the poi made from the taro root, in the same way as a mother would her child.

The dogs are in general useless for anything but being eaten, and are seldom ever heard to bark. In size they are small, with long bodies and ears, sharp pricked noses and short feet. They are mostly black in colour, but very often hairless from mange. In this state the natives feed them and preserve their offspring, which they carry in their arms oftener than they do their own young. In passing their huts, should a stranger hurt one of their dogs, they are more offended than if one of their own children had been attacked.

Meeting To Consider New Laws

At 2 p.m., a meeting was held at Pitt's house, of Lord Byron and all the chiefs, to consider fit laws to be established throughout the islands. Boki stood up for half an hour and made a surprising speech. His own countrymen were so astonished, they said it must be someone else in Boki's skin. He compared what he had seen in England with his own country, and strongly recommended them to establish laws and religion on the English principles. He wished fires to be allowed in their houses to cook on Sunday, and also to be allowed to bathe on Sundays as in England. After discussion, however, it was resolved that the former laws of Tamahamaah should be again put in force.

Marines Too Drunk To Drill

At 3 p.m., the marines came on shore to go through movements for Pitt's gratification, but unfortunately several got intoxicated on their way through

BOKI and LILIHA
From drawing by
Sir George Hayter, P.R.A.

town and could not go through the affair to Lord Byron's disappointment, for most of the royal family and the chiefs and European residents were present.[43]

Wilkinson. Novara's Inn

Wilkinson, now much better, was there, and was anxious to get Lord Byron to obtain a promise from Pitt that the land given him should not be taken from him, but this Pitt and the chiefs will not promise, only saying "He could have it until the ground may be wanted by themselves."

I stayed the night in the inn of Novara,[44] in company with the assistant surgeon.

Queen Kaahumanu, John Young and Chiefs Sail on the Blonde

June 7. Lord Byron informed me I need not send my empty plant boxes on board, as he intended to touch here again. Queen Kaumanna and Pio, with Mr. Young and several chiefs and other natives came on board to go with us to Heddo Bay, situate on the east side of Owhyee.

Took leave of Mr. Charlton, with hopes of seeing him again at Otahite on our way home. Got under weigh at 3 p.m. Fresh breeze from E.N.E. Among the queen's attendants was an old cunning fellow, "Jack Bligh," native of Otahite, who spoke a little English, and had, he said, been with Captain Bligh in the *Bounty* at the time of the mutiny. We had on board also two missionaries and "Sir Joseph Banks," interpreter to Lord Byron.

June 8. Fine. Fresh breezes from the N.E. At noon the island of Mowee in sight. Therm. 6 a.m. 76 ; 12 noon 75 ; 7 p.m. 76.

June 9. Being near the island of Mowee, a boat was sent on shore with Joe Banks, to procure fish. He returned with a quantity, fresh and salted,

43. History repeats itself. In 1825, the British marines apparently got pretty well loaded up with okolehao (native gin). Nearly a century afterwards, the same scene was being enacted in Honolulu. According to a Honolulu newspaper of April 4, 1921, "Fifty-two (U.S.) soldiers were brought to the police station, Honolulu, Saturday night and early Sunday morning. This establishes a record for a single night's arrest of soldiers. They were brought in individually and in groups, practically all being under the influence of liquor, according to the police." This happened in "dry" America.

44. Novara's inn was situated on the "mauka" side of what is now called Merchant Street, nearly opposite the present police station. Joseph Novara died at Honolulu, 14th February, 1839, aged 72. Was a sailor by profession, and had lived in Honolulu for over 30 years.

which the natives immediately began to devour raw, with some poi and taro, not even wasting their gills or entrails. These parts are considered by them to be the most delicate parts of the fish, and were what was generally eaten by the queens and chiefs. Among others doing this was Manaware,[45] whom we had brought out from England. When I, disgusted, expostulated and told him he ought to have shown his countrymen how he had seen us eat our meals, he replied, somewhat offended, that he and his countrymen liked fish in that way best, adding that "He saw plenty of poor people in England, but we see none here; that they got plenty of poi, taro and fish and no want for anything like many a man at home." Such was the answer of a person who had been in England with their king, and who, on the passage back to their own country, had been fed by Lord Byron in a style not inferior to what he usually did himself.

June 10. 7 a.m. Saw Mouna Kaah far above the clouds, in places covered with snow. Fine appearance. Light rain after dark.

June 11. Fine strong E. breezes. Mouna Koa, with its round flat top in sight, bearing South, and Mouna Kaah N.E., as also the high land of Mowee about N. During the day, several showers, fresh breezes and cloudy at 7 p.m.

Arrives at Hilo. Prepares For Ascent of Mauna Kea

June 12. Sunday. Strong E.N.E. breezes and cloudy. At 10 a.m., church service, the queens, chiefs and missionaries present. Shortened sail and came to anchor in 6 fathoms. I got Lord Byron to gain Queen Kaumanna's consent for me to have 7 or 8 natives to accompany me to Mouna Kaah. After her usual "hesitation to consider," she said I might have as many as I wanted. I also asked her for a hut on shore to which to remove my traps tomorrow, where Mr. Forder will live till I return and where he can dry what plants I may find necessary to send home while on my journey. She desired that I should be informed that she did not know of a hut, but when she went on shore she would enquire of the chiefs.

Rev. Mr. Goodrich, Missionary

June 13. Went on shore to find the huts of the only two foreigners at this place, besides the missionaries, to procure one of these men as a guide to Mouna

45. Manuia.

Kaah. I met Mr. Goodrich, one of the missionaries from Woahoo, who told me that both of the persons of whom I was in search had left the place a fortnight ago, to kill wild cattle near Mouna Kaah, and would probably not return for some weeks. He said that rather than I should be disappointed, he would willingly accompany me. His kind offer I accepted.

It was thought best to go the first part of the journey by canoe, and so save 30 miles of travel over many deep ravines and large rivers. We might return by land if we wished. For this water plan we had again to apply, through Lord Byron, to Queen Kaumanna for a canoe and also extra natives to man it. This Lord Byron, in his usual pleasant manner, promised to do when he found her (Queen Kaumanna) in such humour as likely not to refuse him, she at present being rather sulky from accounts received of some persons on shore having acted wrongly in her absence.

Lord Byron gave Mr. Talbot, fourth lieutenant, and Mr. Wilson, purser, permission to accompany me on my journey, and also acquainted me that Queen Kaumanna had promised me the canoe and natives for the next day. At noon I went on shore to choose a suitable hut, and met Mr. Goodrich, who went with me to look at the huts round the bay, all pleasantly situated under the shade of breadfruit trees, which in places form woods by themselves, and grow to a great height, producing plenty of fruit, although they possess but little variety and are generally of the small kind. There are here also plenty of both kinds of rose apple and coconut trees, some of the latter being of great height and age.

Heddo Bay and its neighbourhood are very pleasant on account of its woody appearance and plentiful supply of water, and is far superior in scenery to any spot we had seen either at Mowee or Woahoo, but like these is little cultivated, and only round about the native huts with patches of bananas, taro, tapa plants, etc. At the west end of the bay is a fair sized river with several waterfalls, convenient for watering ships, and in the centre of the bay is another subterranean one, which is dammed up a little distance from the sea and forms a large fresh water fish pond, tabooed for the use of the king and the chiefs. Toward the east end, not far from the shore and near a reef which runs partly across the bay, there is a small island thickly covered with low coco-nut trees, from which circumstance it is called Coco-nut Island.

The whole of the E. side of Owhyee, which is divided into two districts, belongs to Kaumanna and Pio. When at Heddo, their place of residence to receive the rents, is near the east side of the bay, and consists of no more than two huts, one of which is given to Lord Byron as a residence while here.

Returning on board, I heard that the canoe and natives would not be ready until tomorrow. Mr. Young this evening gave me some account of Mr. Menzies' journey to Mouna Roa,[46] next highest to Mouna Kaah to which I am going. During the 36 years that Mr. Young has been on the island, he has never seen Mouna Kaah free from snow, but has not seen snow on Mouna Roa in summer, and on this he bases his theory of the greater height of Mouna Kaah.

June 14. Went on shore with my traps, taking Mantle and another lad Trounce with me. They both belonged to the ship, and are allowed to me as long as I need them. I found that the hut promised me by Manaware was now refused, and only part of another offered, at the other side of the bay, and inhabited by a chief. My traps and provisions now being landed on the beach and surrounded by crowds of natives who would not have hesitated to make free with what they could lay hold of, I begged to be allowed to put them in a corner of his Lordship's house. Lord Byron told me it would make no difference to him leaving any of my things there if I liked till I returned, but if I wanted a place for them and Mr. Forder, I could have the tent put up near his hut for his servants, and this I accepted.

I went with some of the missionaries to Queen Kaumanna's hut to ask her whether I could depend upon the canoe for tomorrow. I found her, as usual, lying on the floor with her face downwards, and several natives round her brushing the flies away from her body. She hesitated in giving an answer until she had surveyed me from head to foot, and then said when she saw one of the chiefs, she would let me know. So I got Mr. Young, who had more influence with her than the missionaries, to tell her I would pay what money she wanted. This offer had the desired effect, for she instantly sent across the bay to the head chief, and when he came it was settled at once that I should have the canoe and natives without paying at all. I sent word to Talbot and Wilson to have everything ready on board for the morrow's start when I came alongside in the canoe. I dined at 4 p.m. with Lord Byron, the surgeon, chaplain and painter, who are his usual companions while on shore. Mr. Forder joined me at sunset, and we took up our abode in Lord Byron's servants' tent.

---

46. Archibald Menzies, a Scottish surgeon and naturalist, was the first white man to ascend to the top of Hualalai and the first white man and probably the first human being to reach the summit of Mauna Loa. For an account of his trips up these mountains, see *Hawaii Nei 128 Years Ago*, Honolulu 1920.

Start For Mauna Kea

June 15. Fine day after a showery night during which the rain poured through the old tent. Mr. Goodrich arrived at daylight with the double canoe and natives, and we immediately began to embark our provisions, etc., for our journey. It was 6 o'clock, however, before we got alongside the ship, for Messrs. Talbot and Wilson, who were ready waiting for us. There were now 17 on board the canoe, eleven natives and six of ourselves. We started with the well wishes of all on board the *Blonde* for our journey of 30 miles to Lapahoi[47] on the W. side of the island.

Favourable light east breezes, which freshened every hour until we landed in a narrow creek at 11 o'clock a.m. The creek was full of rocks, and open to a high surf that is generally found on this coast, and which at all times, except early in the morning, makes landing very difficult and dangerous, as we ourselves experienced. We had the greatest difficulty to prevent our canoe from being dashed on shore, owing to the surf washing over us every minute and filling the canoe with water so fast as to render our efforts in baling it out useless. We got into dry clothes as far as possible and dried our firearms, and then found that the 40 lbs. of salt meat which I had for my share of the provisions was missing, but nothing else.

Laupahoehoe

Lapahoi is a small stony flat with a few huts and sweet potatoes and taro patches scattered over it. It lies at the extremity of a deep ravine, the declivities on either side nearly 500 feet in height and extending to the sea beach, terminating in a rocky precipice. The coast all the way to Lapahoi was intersected by many deep ravines, many of which had large rivers forming beautiful waterfalls that fell over the outward cliffs into the ocean, the angry surf of which broke a long way up upon the rocks underneath.

On the upper part of the inclines a species of pandanus grew plentifully. It is commonly used by the natives for making mats for the floors of their huts. It forms thick plantations here, giving the coast a pleasant appearance with their green bushy tops hanging pendant over the rocks where underneath in many places small subterranean streams fall down at no great distance from each other. This species of pandanus is nowhere so plentiful in

---

47. Laupahoehoe.

the Sandwich Islands as on the island of Owhyee. It is cultivated elsewhere frequently for its leaves for mats and pillows for the natives. The tea tree is also plentiful here in the valleys along the coast.

Climbing Mauna Kea

By noon we had finished taking some refreshments and dividing our baggage into loads for the natives to carry. We proceeded on our journey, leaving behind us six natives with orders to remain four days with the canoe in case we might return in that time and select to go home by water. The other five we took with us, making with ourselves eleven. On the summit above Lapahoi, we stopped to draw breath, and then every step became more interesting as we followed the narrow path to the woods above, which were yet four miles away. As we went along, the few native huts on either side were fast disappearing. The whole face of the country from the coast to six miles inland produced various fine prospects which reminded us of home, and if only cultivated, would produce an equal return of crops to any land of similar climate. But it is not even pastured by live stock, being covered with long grass and short stumpy tree ferns belonging to the *Cyathea* tribe, whose roots afford food for the swine about the huts of the natives.

These same huts are often inhabited by four generations, huddled together at night time like so many dumb animals, and often without sufficient shelter over them to protect them from the cold heavy dews that invariably fall here at night. We reached the outskirts of the woods between three and four in the afternoon, having on our way crossed three narrow deep ravines, thickly covered with wood, mostly *Metrosideros*, *Aleurites*, and a species of *Rhus*, but without water except during heavy rains.

Our guide (Mr. Goodrich) recommended us to take up our quarters in these huts for the night, as these were the last inhabited ones on our way to the mountain where we had any chance to procure food to eat now and also to take with us, which on account of our loss on landing in the surf, we should now need. When about to enter the largest of the huts to prospect its condition, Mr. Goodrich was accosted by a smiling young woman, the wife of one of those Europeans who had come to kill wild cattle. She informed us that she had only left the Europeans yesterday morning, and that they had shot two bullocks the day before. We went and took possession of the cleanest part of the hut for our accommodation, without leave, as is customary with these people themselves, while Mr. Goodrich went in search of a young pig

RUINS OF JOHN YOUNG'S HOUSE, KAWAIHAE, HAWAII
W. F. Wilson, photo.

or fowls. All that he could procure, in spite of offering money and looking glasses, were a couple of fowls, owing to the price put upon their pigs, being nearly treble their worth.

Mr. Wilson was found in the midst of a crowd of natives, highly amused and viewing them with surprise. I went to the wood, while supper was being prepared, to look for plants, and found several species of ferns not seen before, and a few plants. I only got as far as the outskirts of the wood and the trees, which were of moderate size, consisted mostly of *Metrosideros* and *Aleurites*, with many ferns growing beneath their shade. In addition to the different species of *Metrosideros* in variety of colours of the flowers as well as foliage already met with at Woahoo, there still appears in this island many which will add to their number, one particularly with straw-coloured flowers and white underneath the leaves, met with this evening, although sparingly, adds to my former collection.

When I got back, I found my three fellow travellers sitting on a mat, each holding a piece of fowl in one hand a clasp knife in the other, busy eating in the presence of a number of natives, some of whom had in their hands a light made from the kernels of the kukui or candle nut tree (*Aleurites*) several nuts being passed through on a splinter of bamboo cane which gave a greater light than two or three common sized candles.

At 9 p.m. we retired to rest in a corner of the hut on a clean mat brought with us for the purpose, the rest of the hut being filled with the usual medley of men, women, children and dogs.

June 16. Fine but somewhat foggy. Got up at daylight, took the temperature of the air, which stood at 64. We were all ready to start at 5 a.m. in spite of the heavy dew which was still on the grass and bushes, and we were soon wet through by it up to our knees. We entered the wood about a mile from the edge of a small ravine, by a narrow path, where on either side grew a number of strong, healthy banana trees without cultivation and many of them having large bunches of fruit.

John Young and Isaac Davis's First Battle

Mr. Goodrich informed us that it was at this ravine that Mr. Young and Mr. Davis had fought their first battle in the service of Tamahamaah and defeated upwards of 10,000 of the enemy with only 300 on their own side, before their leader came up to their assistance with the main body of the army. The description related to us of this engagement was that when King

Tamahamaah had conquered the south side of Owhyee, he soon after, with his army, marched round to the opposite side of the island by the east, taking with him Young and Davis for the first time, to whom he gave command of the chief part of his army. The chief of the Heddo part of the island was prepared to meet Tamahamaah in order to defend his proportion of the island from being subjected to the other's power, but on seeing the superior force of Tamahamaah, this chief kept retreating to the west till overtaken by Young and Davis, who were nearly a day's journey in advance of the main body of the army. The attack took place early in the afternoon from the opposite sides of the ravine in the wood, when after several hours engagement, it was decided in favor of Young and Davis, who alone had firearms. These two killed the enemy in vast numbers from the crowded manner in which they stood to oppose them, being unacquainted with the destructive effects of firearms.

This battle gave Tamahamaah the conquest of Owhyee.

We halted at 9 a.m. for refreshment, having travelled four miles through the wood, and I had the opportunity to ramble a little out of the path while the others rested. The trees now became more lofty, particularly a species of *Acacia* used by the natives for canoes. Ferns of all kinds and sizes covered the ground beneath the trees, and a good many grew as parasites on the tree trunks. A noble species of *Cyathea*, equally numerous with the rest, often attained the height of 25 feet. *Metrosideros* with red bunchy flowering tops, covered with many red birds sucking their blossoms, were here much larger and taller than any seen on Woahoo. *Beslerias* of various coloured flowers, and some of a climbing nature, and a numerous tribe of *Psychotrias*, both shrubby and succulent, as also many *Lobelias* and other plants, aided by their variety to enliven our journey in spite of the many difficulties encountered from trees fallen across the path every other short distance, that had to be scrambled over. The path being slippery from the night rains occasioned many falls.

Wild Raspberries and Strawberries Plentiful

After travelling another nine miles, we halted to fill our calabashes, this being the last place where we could obtain water till our return from Mouna Kaah. Here again, I took the temperature of the air. It had risen to 69. Towards the end of the wood the path became steeper. Here we found raspberries and strawberries of various kinds covered with fruit which we all ate eagerly to quench our thirst. The raspberries were very large and flat at both ends, but round in the middle and not unpleasant in flavour. The strawberries were small and great quantities of fruit grew around us on every side and looked like a neglected garden.

Bullock Hunters

We reached the end of the wood by 1 p.m., having travelled twelve miles, and above 12,000 feet above sea level. Here we found the two Europeans' temporary hut. They had been killing some of the wild cattle that had originally been introduced by Capt. Vancouver from the N.W. Coast of America and since suffered to remain unmolested for over 20 years. Since the death of King Tamahamaah the government has killed and salted many of the cattle for the supply of its small fleet. In the hut we found both the Europeans at home, asleep, and dressed in the costume of the country. There were also twenty natives, men, women and children outside, some asleep and others roasting pieces of flesh on a stick stuck in the ground slanting over the fire. Both the white men were well known to our guide, and being told of the object of our visit, offered to supply us with what beef we wanted. While the natives were cooking food for us we learned from these two half-naked foreigners, who could speak but little English, although one was a Welshman and the other a Prussian blacksmith, and both for some time had been in the English navy, that they had succeeded in shooting several cattle, but with some difficulty, for the cattle often in droves of twenty were always sensible of any person approaching them. If unsuccessful in killing them with the first shot, it was absolutely necessary to have a place of retreat for their own safety, as they invariably pursued their destroyers with a kind of furious madness while they appear in sight.

Two days before, they had killed an old black bull, which they thought was one of the original number brought from California by Vancouver, from part of the right ear being cut off for a mark. They had been told that this had been put upon the cattle when landed thirty years ago. They have now increased to some hundreds, but it is curious that they have never been seen more than a few miles downwards in the wood from the mountain, and then only in warm weather for the sake of shade and water. Neither has a young one ever been got hold of and domesticated, although often attempted, for the mother living with her young, always seeks some retired place till the young ones are old enough to protect themselves.

I placed all the specimens I had collected since the commencement of our journey, in paper to be left till my return, and then went into the wood to look for more. Took the temperature of the air at 3 p.m., and found it was at 69, being the same as at 10 a.m. coming through the wood. Our guide told us we must travel at least 6 miles further towards the mountain to be able to gain the summit at an early hour tomorrow, before the horizon rose to prevent us from seeing the ship at anchor in the harbour. So waking my sleeping companions, we started on our next stage. However, a native unfortunately dropped a calabash of strong brandy and water (two gallons) being the last of my share of the spirits brought on the journey. We had scarcely travelled three miles when a thick fog commenced to roll in over the country which was covered with tufts of dry grass and full of cattle tracks. The soil was chiefly composed of sandy, pulverized lava, with numerous beds of strawberries growing on same. Raspberries grew in great abundance by the sides of the small ravines made by the torrents of water from the melted snow running here at certain seasons. They were of a better flavour than those in the upper part of the wood, being here more exposed to the sun.

By 6 p.m. we had travelled another two miles, when the fog became so thick that we were scarcely able to see ten yards ahead of us, and we were drenched and shivering with cold and almost beyond any vegetation to shelter us for the night. So we cut down boughs of *Acacia* and a species of *Sophora* and erected a hut. This we accomplished in little more than half an hour, and getting plenty of firewood kept a fire burning all night near where we lay. I rambled about till dark among cranberry bushes cutting specimens. The temperature at 7 p.m. was 52.

Too Cold For Natives

Got up at 2 a.m., started at 3 and began our journey to the mountains leaving the natives behind, who feared the cold and did not want to accompany us. At 5 a.m., daylight began to appear and by then we had travelled three miles over sandy pulverized lava, sinking over our ankles at every step.

The Silver Sword Plant

The last mile was destitute of vegetation except one plant of the *Syginesia* tribe, in growth much like a Yucca, with sharp pointed silver coloured leaves and green upright spike of three or four feet producing pendulous branches with brown flowers, truly superb, and almost worth the journey of coming here to see it on purpose.[48] The majestic clouds rising on the horizon at daybreak encircled us all round like an immense wall with towers of various forms and sizes on their tops. They lay at unequal distances along the horizon, gradually rising and changing into fresh shapes at every moment that had the finest effect imaginable.

Talbot and Wilson Unable To Proceed

The temperature had now fallen to freezing point. Messrs. Talbot and Wilson, overcome by the cold, became so sleepy as to be unable to proceed. We waited by them for some time trying to rouse them without avail, so leaving one of the lads with them, my guide and I with the other lad started out afresh so as to reach the summit and see the ship, we having promised Lord Byron to light a fire that he might see through his glasses how far we had got. As we advanced, every step became steeper and more difficult. All vegetation had ceased, even the yucca-looking plant, but we got up the mountain by 6 a.m., and saw the ship looking to us down there like a 50-ton vessel. Here we collected enough stumps and leaves to light a fire, remaining by it for half an hour, and our companions not overtaking us, we kept on our way, at times over hard uneven lava, at others over sandy lava.

Reaches Main Plateau of Mauna Kea

The mountain now became divided into several high conical sandy hills with several old small volcanic craters on their sides, forced above the sand for some yards in height and bleached nearly white from long exposure. The air became warmer and more pleasant as the sun rose above the horizon, but we had constantly to rest from the difficulty of breathing after stopping to rest. At 8 a.m. we saw the lad, left with Talbot and Wilson, coming after us. Thinking he might have a message from them, we waited, but he had left the others still asleep, and only came to beg to be allowed to return, as he had

48. The silver sword plant of Hawaii was first brought to the knowledge of the botanical world by Dr. W.J. Hooker, who described it from specimens collected by David Douglas when he ascended Mauna Kea in 1834. See *David Douglas, Botanist at Hawaii*, Honolulu 1919. Macrae climbed Mauna Kea in 1825, i.e., nine years ahead of Douglas, and must be credited with having been the first botanist to notice and collect the silver sword plant. Some modern writers persist in stating that the silver sword plant is found on Maui only. This is not the case, as it also grows on the high mountains of Hawaii.

been so cold waiting by the sleeping men. Giving him some refreshment and spirits, we sent him back to try and meet Talbot and Wilson. The temperature had now risen to 46, the sun shining brightly. We resumed our journey by the bottom of the sandy conical hills, the surface over which we travelled constantly changing and more uneven, sometimes being lava sand intermixed with small broken stones about the size of brickbats, and at other places having to scramble over large sharp-edged granite stones of several tons weight, which have beyond a doubt, been thrown up by some previous convulsion. We came in sight of the snow after 11 a.m. Our guide seemed to suffer more than the lad and myself from headache and inclination to vomit, and we had yet two miles to go over a still more rugged surface to reach the snow.

Mr. Goodrich Collapses

At a quarter mile from the summit where the snow lay, our guide collapsed and begged us to get him some snow for his thirst.[49] The lad Mantle held out better than I had expected.

Macrae and Boy Mantle Reach Summit

At 12:30 I reached the snow on the summit, which lay on porous lava of a sponge color, and in places on sand of a red color intermixed with red and black cinders like the conical hills we had passed. Some of these cinders had common quartz and two or three other kinds of minerals that I am as yet unacquainted with, very often bedded in one lump of lava. The snow in some parts was about three feet deep, congealed into solid ice, excepting from two

49. Rev. Joseph Goodrich, who, on this occasion, was unfortunately laid up with mountain sickness, had on 26th August, 1823, reached the summit of Mauna Kea. This is the first recorded instance of the ascent of this mountain, although Mr. Goodrich mentions that on reaching the top of one of the terminal cones that encircle the main plateau of Mauna Kea, he discovered a heap of stones, probably erected by some former visitor. Who this former visitor was is unknown, but he was probably one of the white men that in the early years of the nineteenth century got a living by shooting wild bullocks that roved on the side of Mauna Kea. It is very unlikely that any native had reached the top of the terminal cones on the summit, owing to being unprovided with warm clothing to resist the great cold and also to the fact that the natives had a superstitious dread of the mountain spirits or gods. About six months after the date of the first ascent of Mauna Kea by Mr. Goodrich, the peak was scaled by Dr. Abraham Blatchley and Mr. Samuel Ruggles, both connected with the American Mission.

to three inches at top of rough particles of loose snow. The whole appeared to be melting fast through the porous lava like a sieve, which prevented our being able to fill our vessels with water. We therefore filled our handkerchiefs with snow, taking mouthfuls at the same time to quench our thirst. I sent Mantle with some snow to our guide, and remained to take the temperature in the sun and in the shade. In the former it was 52, and in the latter, in holes beneath the snow, 44. I stayed about an hour admiring the scenery. For a space of about 12 miles around the top of the mountain, it was dreary to a degree, but below that, the pasture where the wild cattle fed had a pleasing effect. The forest which encircles the island of Owhyee below the pasture land, was hidden in fog, so that I only saw about 20 miles in a direct line, but the high land like Mouna Roa and other hills could be easily distinguished above the fog, although none of them were covered with snow.

Sheep Killed Off By Wild Dogs

I saw many skeletons of some kind of animal, devoid of all flesh, but apparently not long dead, and on rejoining our guide, was informed that the wild dogs had almost exterminated the sheep that Vancouver had brought with the cattle, pursuing them beyond the line of vegetation, where they became bewildered and died for want of food.

Begins Return Journey

It being now after 2 p.m., and still feeling unwell from the same causes as our guide, we left this interesting place and travelled slowly downwards, finding our few specimens of minerals, etc., almost twice their real weight. In this hobbling manner, scarcely able to drag our limbs for the last four miles, we reached our hut, where we found that the lad sent back in the morning to Messrs. Talbot and Wilson had not met them. So fearing they might have succumbed to the cold in their sleep, and knowing they had no provisions, we much repented having left them; but to our joy, in about half an hour, we heard them calling not far distant. When they came to the hut they did not appear so fatigued as we ourselves, in spite of the want of food. They had slept for about an hour, then awoke and tried to follow us, but not finding any of our tracks, they gave up the idea of following us, and made for the first of the highest hills. The snow we had brought with us served us well with water, for the natives left behind had drunk all that we left of the latter article except about a pint. The natives rubbed our thighs and legs for us, a practice they often do for themselves in such circumstances. They call it lummi lummi. The temperature at 7 p.m. was 50 and at 10 p.m. 48.

We calculated the summit of Mouna Kaah from Byron's (or Heddo) Bay to be about 70 miles by the common path, but in a direct line perhaps only half that distance. We judged the peak could not be under 18,000 feet above sea level.[50] The land along the sea coast from Byron's Bay to upwards of 40 miles to the west and about 6 miles in breadth, was free from wood excepting by the sides and bottoms of the ravines. The forest that surrounds the central part of the island begins here, at the distance of 5 or 6 miles from the coast, and stretches back for a depth of 12 miles, intersected with deep valleys and large rivers of fine water. The outskirts of the forest nearest the sea are chiefly handsome coloured flowering species which entirely disappear after 5 or 6 miles towards the centre of the wood. The commonest species of *Metrosideros* often attains a height of 40 feet and are thick in proportion. The wood is hard and durable.

The upper parts above the forest resemble pasture land for 7 miles farther, and are thinly covered with low growing shrubs and abundance of strawberries and raspberries. At a higher elevation, vegetation ceases for the last eight miles towards the summit. The clouds generally rise on the mountains of Owhyee and the other islands in the morning and disperse towards evening. Rain often falls at night and also in the daytime some distance from the peaks, while on the coast the sun may be shining and there is no appearance of rain.

June 18. Got up at daylight, being disturbed in the night by the howling of wild dogs which caused us to keep our fire burning. At six set out on our homeward way, and unknown to us, the natives at once set fire to our discarded hut, a common custom our guide told us. At 10 we had travelled 6 miles and reached the Prussian and Welshman's hut. These men had seen no wild cattle since we left them, the only animals observed having been a wild dog and cat. The dog seemed to be the same kind as the domesticated native one of which they eat the flesh, and the cat appeared like the European breed. After a breakfast of plenty of slices of roast beef and abundance of water, my companions spread their mats in the shade and slept till noon. I shifted my specimens that had been left here into dry papers, and gathered others, including strawberry and raspberry plants to take with me to England. At 2 the whole camp was on the move for Lapahoi, where we had left the canoe and the natives. On reaching the first hut, we found only the two foreigners,

50. Mauna Kea is 13,825 feet high.

the rest having gone on to Lapahoi. They promised to have a fire ready for us to dry our clothes at, but although I gave them each a dollar on starting they had nothing ready for us and did not get us any food till 9 p.m.

Natives Object to Sunday Travel

June 19. Hazy, light showers. Sunday, and on that account the natives refused to accompany me to join the other part of my party with their loads, and said the missionaries had ordered them not to.

Game of Noa

The blacksmith, however, promised to accompany me with his own people at 11 a.m., but instead of doing so, went and played and gambled at Nooah. This game is one of their most ancient and frequently played pastimes. It consists in placing in a row, five small tapa bags stuffed with cotton or the down of ferns, underneath one of which is hidden a stone so as to deceive the parties playing which of the bags it was put under. The players are seated around in a circle, each armed with a small wand in his hand with which he strikes the bag he supposes the stone to be under. There are generally ten players with different coloured rags tied to their wands. I have been told that at this game they gamble their hogs and all their possessions, even their wives, and are very strict in paying their debts of honor.

Leaving them gambling, I left with the two lads and two natives for Lapahoi, and joined my party there about 3 p.m. They were just about to start for home, having expected me in the morning. The canoe had not waited for us as ordered but had returned home the second day after we started for the mountain. We therefore had to go home by land, and took up our quarters for the night about 7 p.m. after having crossed a number of deep ravines, wading through rivers, at times up to our middle.

Having travelled about 10 miles east of Lapahoi, the natives lighted a fire, for we were drenched, and cooked some bananas and breadfruit. To do this, they first of all got some stones heated red hot, then afterwards put them into a gourd calabash with a few leaves of the tea tree underneath and the same again above. Then were added cabbages cut in pieces, intermixed with bananas that had only the outer skin taken off. Over all on top were put more leaves to prevent the steam rising upwards. In about a quarter of an hour the mess was ready to eat, and our vegetable supper was by no means unpalatable.

Difficult Travelling Over Ravines

June 20. Cloudy and cold. Got up at 4 a.m., and commenced our journey of crossing ravines and fording rivers, taking off our shoes for surer footing. At one river Mr. Talbot nearly came to grief, for insisting on crossing at the same place as the natives, in spite of their protests that he did not know, as they did, where to find the stones on which to place his feet, after a few steps he was swept off his feet, and managed to get hold of a stone that was above the water whereon happened to stand two natives, who caught hold of his hands, but a third native attempting to lay hold of his body was carried away by the current to some distance not far off from a waterfall. The natives holding Talbot could not drag him out, for one of his legs was caught between two stones. I saw the whole affair but could not get to him, having waited to go higher up to a surer place with Wilson, but I managed to get near enough to get the barrel of my gun under his leg and succeeded in extricating it, after which the natives were able to pull him out of the water.

The natives had shown great anxiety to get Talbot out of danger, and as a mark of gratitude, he gave them each a knife. He was little the worse except for bruises. The natives said that if he had been drowned, Kaumanna would have been sure to have killed them for not saving his life. All the rest got safely across, but we soon came to another river as large, and at this one Mr. Wilson fell and broke our flask with the little remains of our brandy! At 9 a.m. we halted at a large hut and breakfasted off a few taros and some poi. Then continuing on our way we crossed ravines 400 to 500 feet deep with fine, clear water at the bottom. The sides of these ravines had plenty of rose apple, breadfruit trees, pandanus, ferns and other plants.

Gets Back To Hilo

We arrived opposite the ship at 5 p.m. very tired from our many climbs up and down, since we left Lapahoi on Sunday, distant 40 miles.

Mantle who had gone with me to the snow and two of the natives were the only three who had managed to keep up with us all the way. On getting on board the ship, where I was invited to dine with Talbot and Wilson, I heard that Mr. Forder was still with my things in the old tent, so I slipped away, got into a canoe again without dining, to find out why the things had never been moved according to promise.

June 21. Got my specimens into dry paper. The lad Trounce arrived saying he and my things had been upset out of the canoe in crossing the river at the watering place, but as he had nearly been drowned, I could not be angry, so again I had at once to set to work to shift everything into dry paper. Accompanied by Mr. Young, as interpreter, I succeeded in hiring a hut for the remainder of my stay, for which, in preference to money, I gave two blankets for three weeks.

Native Rent Collection

June 22. In the forenoon, over two hundred natives came from the east part of the island with their rents for Queen Kaumanna, such as tapa cloths,[51] and mats of various dyed patterns, hogs, taro, etc., etc. She never attempted to smile or give a pleasant look at the natives who brought this rent, although I heard she had not been on the island during the last four years. I asked her permission, however, and she gave me three of the largest sugar canes as specimens. A mat which was given to the surgeon, I was informed, would take a woman a year to make. When the women get to a certain age, it is their duty to manufacture all the tapa cloths and mats.

On going to my hired hut to take my traps, and offering my two blankets, they did not want to let me have it after all, and the wife and some others shed tears. He told me it was their fear of my burning the hut when I went away, as was their custom. Being assured that I had no such intentions, they at once received me as tenant.

Macrae's Hut Tabooed

The chief man of the place came at Mr. Young's request to taboo my hut, to prevent my being crowded by the natives.

The charm was done thus: The chief stuck in the ground a few wands

---

51. Down to about the end of the fifties of last century, the natives on the island of Oahu were in the habit of paying at least a share of their taxes in the form of tapas. The tapas were deposited in the Pa-auhau or tax yard, situated about where Marin Street is, near the foot of Nuuanu St., and west of Merchant Street. Messrs. Robinson and Lawrence had a ship repairing yard at "the point," near where the fort stood. They were in the habit of buying these tapas from the government and used them for felt under the sheets of copper nailed on the sides of vessels that were being repaired. Tapa or kapa making is now a thing of the past in Hawaii, but the process of making it may still be seen in Fiji, Tonga, and Samoa. In the Cook group, Tahiti and other South Sea islands, tapa has given place to gaudy colored cotton prints, woven in Manchester and dyed in Glasgow.

at some distance from each other, all round the hut, which on being seen by the natives, they never dare enter, unless sent with some message by any of the chiefs.

Botanizes in Puna

June 23. Went about nine miles east of Byron's Bay to collect plants, but met with little success. Messrs. Talbot and Wilson called to say they were going to the volcano on the 25th, and had come to make arrangements as to a guide and provisions to take with us.

Journey To Kilauea Volcano

June 24. Arranged with the blacksmith, met at Mouna Kaah, and five natives, to act as guide and carriers. The former recommended me to speak to Kaumanna in order to get her to order the latter not to leave us till our return.

June 25. Talbot and Wilson and the guide appeared, but none of the natives, so the guide had to go in search of them. By 7 a.m. the guide came back with the natives, and we began our journey as the first party for the volcano from the ship. His Lordship and several others intend to follow us in a few days. We travelled thirteen miles by 1 p.m., five miles of that distance being through a wood over a narrow path of broken pieces of sharp edged lava, which we could hardly bear our own weight upon without pain to our feet. This wood had many trees, mostly *Metrosideros* of over 40 feet, but slender in proportion, and far inferior in size to those we met on our way to Mouna Kaah. Under their shade grow numbers of ferns which hide the lava.

At 6 p.m., having travelled 20 miles since leaving the ship, we reached a hut newly put up for the use of Lord Byron when he passes this way. The last six miles were through an open country, over solid greyish black undulated lava, covered with stumpy ferns, chiefly *Cyathea*, which the natives often burn during the dry season. We found also quantities of three sorts of cranberries on each side of the path. They were ripe and we enjoyed them much although somewhat acid. The road through which we had come, continued all the way up towards Mouna Roa.

We were joined in the hut for the night by numbers of natives, who without the slightest compunction, examined the skin of our hands and feet to see if it were the same colour as our faces. But a look from us would make the children under ten take to their heels. The old women generally nursed the children, carrying them on their backs, never in their arms. When travelling any distance, they tie the children to their back with a bandage of tapa cloth placed round the child's neck and thighs, so as to leave the woman's hands at liberty.

June 26. Wet and foggy, so could not resume our journey till after 8 a.m. Reached the last native huts on the way to the volcano by 12, having travelled 10 miles over the same kind of undulating lava as yesterday. Our guide here spent two hours trying to procure some fowls or a hog to take with us, but the natives wouldn't part with them unless we gave three times their value, so we started again with what little provision we had of our own with us. Our natives grumbled, saying they would have to eat ferns before they got back. However, after we had left the huts some distance, the natives who lived there sent after us a couple of fowls for which they would take no payment.

By 6 p.m. we reached two old hovels on the outskirts of the wood of *Acacia* trees, having travelled about 18 miles since morning over lava covered with ferns, cranberries and low straggling bushes of red flowering *Metrosideros* and along the last two miles, beds of strawberries growing under the ferns and grass on sandy pulverized lava. Here we stopped for the night, our guide saying we were not far from the volcano. Repaired the hovels, the natives making a fire to cook the fowls. Our guide reported he had seen the volcano smoking in the distance, but we hardly believed him.

Kilauea Iki

June 27. Misty and foggy morning, so could not start till 7 a.m., when we passed on our left a large old volcano crater, over 1000 feet deep, now covered with verdure on its internal declivities, and the bottom, which is level, having a few low growing red flowering *Metrosideros* bushes.

Arrives at Kilauea Volcano

At 8 a.m. we reached a shed in good condition, situated on the edge of the active volcano, which we now find we could easily have reached last night instead of staying and repairing the hovels. Besides we could have had the gratification of watching the burning craters during the night. Here we stood gazing on the immense depth below covered with clouds of smoke, while at short intervals a terrific noise was distinctly heard among the different burning craters.

Descends Into the Kilauea Volcano Pit

In the meantime the natives were busy making sandals of grass to protect their feet from the lava when they got below. At 8:30 we all eagerly began to descend from the hut, with walking sticks, there being nothing to hold by but short tufts of dry grass. The loose stones kept rolling amongst us all the way down to the first ledge, which we reached at 9:30, without any injury beyond bruises from the stones. The vegetation ceased at this point, the burning craters being at least 500 feet below us. Our road became more difficult and steep, over large irregular sharp-edged stones torn from their bed, piled up loose upon each other and intermixed with flat pieces of honey-combed lava.

By noon, with difficulty and danger, some of us had reached the nearest smoking pillar, about 30 feet high and covered with sulphur, which gave it a beautiful yellow appearance. We waited here some time for the natives to come up who had hurt their naked feet and legs in falling in through the hollow lava that lay in places resembling flues on the top of the more solid material underneath, which required the greatest caution to try it first with our sticks to see whether it would break before we attempted to advance a step upon it. We crossed many wide vents. Some of these openings were constantly smoking and smelt so strong of brimstone that got up our nostrils when going over them, as nearly to suffocate us.

While standing by the brimstone pillar, we noticed that at times there issued forth sudden gusts of smoke, strongly impregnated with brimstone, which obliged us to be careful to avoid when we approached near it for the purpose of picking up specimens of the hot lava covered with sulphur from the opening in the sides. The natives who had experienced difficulty in coming thus far, besides fear, now begged leave to return again to the top, which we readily granted, while we four went farther on to examine another pillar about 50 feet high, burning red at the top like a furnace and emitting to a considerable distance from it, lumps of soft lava and cinders of various colours. On getting near to the pillar, the terrific noise kept up underneath everywhere round it for a considerable way from where we stood, intimidated us that we were afraid that while we remained there, the part underneath us might be torn to pieces every minute.

Frightful was the noise kept up by the burning beneath among the numerous pillars, which were nearly hidden in smoke and impossible of approach, except to windward, owing to the strong smell of sulphur. The noise of the burning craters resembled a blacksmith's forge, and could be distinctly heard half a mile away. At 1 p.m., as we had not time to stay longer, we began to re-ascend at a point a mile farther to the north of the hut from which we had started in the morning. We found the temperature to be 92 in the shade, and in the sun it exceeded the height of my thermometer, which was only marked up to 132.

The volcano is situated about 40 miles S.E. of Byron's Bay, at the foot of Mouna Roa. According to the missionaries who measured it some years ago, its circumference is 7 miles round the edge at the top, and five miles round the almost circular basin below. The depth is in most places from 1000 to 1200 feet. The ledge, which is nearly perfect all round the basin is rather more than half way down from the top. The number of burning craters, in the form of pillars, exceeded twelve, some of which were over 50 feet in height and generally of a bronze shining colour and of various shapes, with several unequal openings on their sides to the summit, where sudden gusts of smoke issue forth at intervals.

Adjoining the craters where we had travelled, the surface varied. In some places the ground was covered with sharp-edged granite stones, piled up above each other, intermixed with similar hard stones of a red colour, or smooth and white, but in general the basin was composed of black and greyish porous honeycombed brittle lava, often undulated like waves of the sea, and where late convulsions had taken place, it formed at short distances hollow burning passages of great length, which are full of red and white ashes. When the thin crust above these passages are trod upon, it easily falls in. Most of the lava was crystallized in small particles on the top like glass, which if fallen upon, tore the skin of hands and feet through our clothes.

The internal declivities on the west side are steep and free from vegetation, but on the south and east sides vegetation continues in most parts to the first ledge and consists chiefly of the red flowering *Metrosideros*, dwarfed in size, and mixed with tufts of two or three kinds of dry grasses. To the south we noticed a large space covered with brimstone, looking like the chalky cliffs of Dover and Gravesend. At 2 p.m. we gained the summit on the north side, where, near the top, I saw several bushes of sanders wood in flower, with a few other plants, but there were not the same variety as met with when nearly above the zone of vegetation on the slope of Mouna Kaah.

On the flat land at the top were several hot springs underneath narrow deep openings divided in the earth at short distances apart from each other, and spread over for half a mile in breadth from the edge of the volcano top. For more than a foot wide, these openings had on the surface of their sides, luxuriant patches of green moss growing from the moisture caused by the steam. Some of the water near the surface which we drank, was only warm. Here we noticed some sheds used by the natives when cutting trees for canoes. We also saw some remains of cooked fern (*Cyathea*) which our natives are glad to eat, and which we found not at all unpalatable.

We reached our previous night's resting place by 4 p.m., having met on our way with quantities of cranberries and strawberries. Some of the sulphur specimens which I had put in seed paper when down in the volcano at the first pillar, I found had already burnt through the paper and my coat pocket. Its taste was more acid than of sulphur. We then travelled homewards and reached the first huts soon after dark, and found waiting for us, the natives that we had allowed to return from the volcano. They had some taro and poi ready, being all the food they could obtain. Having mixed the poi with a little sugar brought by Talbot and Wilson for the purpose, and warming the taro before the fire, we managed to satisfy our hunger, and then retired to rest, hoping to get home the next morning.

Meets Lord Byron and Party En Route to Volcano

June 28. Got up at daylight and started for home. When we had gone four miles we met Lord Byron's party going to the volcano. It consisted of 30 natives with the head chief (Maro) at Byron's Bay as their leader, who had gone on ahead to have fires lighted to cook a meal for his Lordship. A little after these we met the naturalist and one of the missionaries, followed by the surgeon and draftsman, then the chaplain with two natives carrying his hammock slung to a pole, ready for him to get into when tired. Again after these came the surveyor, first lieutenant and another missionary and a boy middy. A short way in the rear was Lord Byron, accompanied by another middy who had his hammock with him ready to be carried like the chaplain. When we informed Lord Byron of our want of provisions, he told us to take anything we wanted from the carriers, as he had plenty of everything. We reached the hut where we had slept the first night at 8 p.m.

June 29. Sorted and packed up my specimens of lava.

Native Burial Customs

June 30. A man about thirty years of age died in the evening, and I found his relatives had taken him out of his hut to breathe his last in a temporary shed, owing to a superstitious custom among them never to inhabit a hut where anyone had died, but to burn it immediately. The body was lying on a dirty mat rolled up in a piece of tapa cloth, with a few natives making a howling noise over it. On asking them what was the matter, they leaned their head on one hand, while they pointed the other towards the sky where they said he had gone to sleep and parted from them for ever. I told what I had seen to Mr. Young, saying I had never seen a funeral since I had been here. He replied that he had only seen one and that by a great favor. It took place in the dead of night, and was made up of a few old men and women who are always entrusted to bury the dead about that time with great secrecy, for what reasons he did not know, only supposed it to have arisen from an old superstitious custom. He said some of them never buried their dead, but concealed them for a certain time till the flesh decayed off the bones sufficiently dry to be tied up afterwards in a bundle of tapa cloth and then hung up inside the roof of their huts.

July 1. Lord Byron and his party returned from the volcano, some of them very lame.

Visits Waianuenue or Rainbow Falls

July 2. Went at 5 a.m. to see the waterfall above Byron's Bay, with Mr. Talbot and Mr. Goodrich, our Mouna Kaah guide. The fall was situated about six miles inland and was over 60 feet in height. The river is one of the largest on this side of the island, and like the rest has many falls upon it at no great distance from each other. I remained with Mantle in the wood till dusk, gathering plants, some of which I had not seen before, particularly a beautiful climbing plant, resembling an erythrina, with crimson flowers,[52] and several *Metrosideros* and ferns. I learned on my return that Lord Byron intends to sail on Thursday, the 7th for certain, which prevents me from going again to collect in the woods.

July 3. Sunday. Fine. Busy all day among my specimens and seeds and writing. Another party of gentlemen from the ship have gone to the volcano to be back before the ship sails.

July 5. My last night on shore at Byron's Bay, as I have to be on the ship with all my traps by tomorrow morning.

July 6. At 4 a.m. a boat came and took me with all my belongings on board.

---

52. Nukuiwi (*Strongylodon lucidum* Seem.), i.e., Iiwi's bill or beak from the shape of the flower.

NUKU'IWI
"Iiwi's bill" (*Strongylodon lucidum*)

Leaves Hilo For Oahu

July 7. Weighed anchor at 5 a.m. and sailed for Woahoo taking with us the queens we had brought from thence, also two missionaries, their wives and families and over 100 natives. At 2 p.m. still light, variable winds, so lowered the boats to tow the ship off the land.

July 8. Fine, with fresh breezes from the east. Made Diamond point, near Hanarura at Woahoo. Stood off and on the land for the night.

Honolulu Once More

July 9. Fine. Moderate breeze from the N.E. Came to anchor off the harbour of Hanarura. The pilot Adams came on board and reported that the *Asia*, 64, a Spanish ship, with two brigs in company, who were pirates on the coast of California, were said to intend touching here for supplies. This information came by a schooner lately arrived from the coast. In consequence, all was soon bustle to get the natives and their luggage on shore as soon as possible so that the guns might be mounted ready for action. At 9 a.m. Lord Byron requested me to go ashore and get my plant boxes on board at once, as he might get under weigh this evening and join the schooner at Karakaakua Bay. Went and returned with the boxes and found all the natives gone and the guns mounted and loaded, the sailors in high spirits, hoping for the enemy to give them battle and so gain a prize.

Marin, the Spaniard

July 10. Sunday. Fresh N.E. wind. Went on shore as we were not to sail till tomorrow morning, not to return here again. I went to see what plants were left of those brought out from England. Met Lord Byron and Marin the Spaniard. We found only six plants and these without leaves. Marin said he had removed the others to his farm in the mountains, where many were doing well.

To Mr. Marin's diligence as a farmer, the Sandwich Islands, particularly Woahoo, may be said to be under some obligations for his introducing and multi-

HONOLULU
(Looking towards Barber's Point)
Drawn by L. Choris.

plying various kinds of animals, which in a few years will be plentiful on the island. When I was about to leave Mr. Pitt, I requested the favour of Marin to inquire of him if he would give me some patterns of their native cloth for the Horticultural Society, and received for answer that he would consider of it. Remained on shore with one of the gentlemen from the ship at the inn of Novara.

Dusty Honolulu

July 11. Fine, with strong wind, which raised clouds of dust in the town that nearly suffocated and blinded me. Since we had been at Owhyee, I found that Pitt, who was 60 years of age, had married a young woman of 19. She, with her sister, had lived with him for many years. This marriage is the second which has ever yet taken place in these islands. Went on board where all was in readiness to get under weigh when Lord Byron came on board.

July 12. Lord Byron on leaving the shore to embark, was saluted with 13 guns from the forts and was accompanied by Pitt, Boki and several other chiefs to take farewell of us. It was now understood on board that we were to leave the islands for Otahiti, after touching at Karakaakua Bay to take on board the surveyors out of the Sandwich Island schooner given them for their use by the government to survey on the coast. Pitt was most interested in examining everything on board, and requested Lord Byron to make all sail and get some distance to sea that he might see how we sailed. To humour him, Lord Byron had all the hands mustered and as they passed along in front of them they all saluted, to the scandal of some of the middies, etc., that British men-o-wars' men should salute a half-civilized savage, unaccustomed to such honour from his own bareheaded countrymen. At 3 p.m., made in again for the land [and] sent them all on shore with a salute of 13 guns from the *Blonde*.

July 13. Fine, with moderate wind. Made sail as yesterday evening for Karakaakua Bay.

Anchors At Kealakekua Bay

July 14. Saw the high land of Owhyee, above the clouds, covered with snow. At 10 a.m. came to anchor at Karakaakua Bay in 26 fathoms. Went on shore and were shown where a cannon shot had gone through the trunk of a coco-nut tree, near where Capt. Cook was killed. Some of the others of our party went close to the water's edge looking at the spot where he fell. This place is at times overflown by the tide at high water. Having satisfied our curiosity here, we went to examine some caves close at hand in front of a steep volcanic rock which divides Karona from Karakaakua.[53] The first cave was full of dozens of old muskets apparently the manufacture of various countries. Scarcely one was complete. They were heaped up, broken in pieces and covered with rust. While hunting in the corner for curiosities, we found an old fishing rod (which was given to me) and a club. The rod was chiefly made of human bone, and the club made of wood shaped like a thigh bone. However, Lord Byron seeing the chief at his hut, prevented me from keeping the rod, as we had no permission to take anything away, so to our disappointment, it had to be replaced where we found it.

We next visited the spot where Captain Cook's body was said to have been divided by the chiefs.[54] It is about half a mile above Karona, on the top of a hill. We found the place had been held sacred for some time, judging from the appearance of the old lava walls raised round about 20 feet square and from 4 to 5 feet high, and nearly the same thickness.

In the evening we went across to the other side of the bay at Karakaakua, which, like Karona,[55] is composed of only a few straggling huts, under coconut trees growing entirely upon lava, without any signs of anything nutritious to support them. Near where the immortal Cook was cut up to be divided among the chiefs, grows a fine white flowering plant of a low shrubby nature not met with before in any other part.[56] It belongs to the tribe *Diadelpha* and is generally found only on the lava near to the sea coast, with scarcely any other vegetation near it. It had plenty of seeds which I gathered and am in hopes that it will grow in England.

July 15. Owing to the uncertainty as to our starting, I could only pick up the few plants growing on the barren fragments of black lava near the sea shore and for several miles inland. Beyond the fringe of bare lava, the ground

53. Kaawaloa from Kealakekua.

54. This spot is known to Hawaiians as "Kapuhiolono," "the burning of Lono," i.e., Captain Cook.

55. Kaawaloa, or as it is generally pronounced Ka-ava-loa. There is a distinct "V" sound in some Hawaiian words, such as "Ewa," pronounced "Eva." Andrews in compiling his dictionary of the Hawaiian language, saw fit to ignore the "V" sound, as he has also that of "T," which is still to be heard on the island of Kauai or Tauai.

56. Here is possibly meant Puapilo (*Capparis sandwichiana* D.C.).

gradually rises to the woody mountains that appear covered with cultivated patches of sweet potatoes, taro, breadfruit and banana trees.

The whole coast on this side of the island has a horrid and dreary appearance. The ground is everywhere covered with cinders and streaks and hollow parts which mark the course of lava from Mouna Roa. The natives have shown great industry in clearing away the top lava to the depth of 4 feet or so in order to get to the more fertile soil underneath, which is composed of rich ashes and a light mould. On Owhyee, the taro is little cultivated in ponds of water as at Woahoo and Mowee, and is consequently of much smaller growth but dry and farinaceous. The sea here, like at other parts of the islands we touched at, abounds with a variety of excellent fish of many colours. What little water is met with is generally brackish.

Cook's Observatory

We saw at Karakaakua the large heap of stones where Capt. Cook had his observatory, which was formerly a morai and religious places for human sacrifices.[57] We also noticed near the same place three coconut trees with cannon shot holes through them from the firing upon the natives which took place when Capt. Cook was killed. In the gardens of the native huts grow a few tobacco plants and banana trees for immediate use. We saw a dog being cooked for one of the chiefs, who consider them better eating than any other animal.

Visits the Puuhonua at Honaunau and Hale O Keawe

July 16. Went to see the morai[58] on the other side of the island. On our way met the old priest in his canoe coming on board. He alone is entrusted to enter the morai, and we accordingly took him back with us. We found the morai was on the east point of a small bay surrounded by huts standing under a thinly scattered grove of coconut trees, but with no signs of cultivation about. As we were about to enter the morai the old priest, who had on a straw hat and cotton shirt, took both of them off, and only left his maro on. On

---

57. Hikiau heiau or temple.

58. By morai is meant the Hale o Keawe. Although Macrae does not mention the large Puuhonua or City of Refuge, he must have visited it as it immediately adjoins the ground where the Hale o Keawe stood. For particulars about the Hale o Keawe, see article by Prof. W. D. Alexander in Vol. 3 of the *Polynesian Society Journal.* Dampier, the *Blonde's* artist, made a sketch of this building, which has often been republished.

entering we only found an empty filthy hut with quantities of human bones in heaps under mats at each end of the hut, many of the bones not yet dry and disgusting to the sight. In the middle were several effigies of the deceased chiefs, tied to a bundle of tapa cloth containing the bones of each person whom the effigies represented. Most of the effigies were made of wood, but the one representing the late Tamahamaah was substituted by a mask of European manufacture and was more finely dressed than the others. The party with Lord Byron that had visited here the day before, had taken away any memorials of the morai that could be taken, so we asked the old priest to be allowed to take some of the ancient weather beaten carved figures outside.

The morai is a small thatched hut fenced round with sticks to the height of 6 feet, kept together by two rows of bars. Fixed in the lava ground at the entrance front stand upright several various sized wooden rudely carved hideous figures, in representation of their former gods. These they now set but little value upon, and are rarely met with in the huts of the natives.

We travelled home by land along the sea coast, over uneven masses of lava, and meeting with only a few plants such as tufts of low-growing euphorbias, convolvuluses, sidas, but mostly diadelphas plant seen common at the place where the chiefs shared between them the body of Capt. Cook. There was also a species of hibiscus, with yellow flowers, much used here by the natives to dye their tapa cloths yellow. On the way home, we passed a number of heaps of lava raised over the dead slain in one of Tamahamaah's battles for the conquest of the island. This was between the morai and Karakaakua.

On board we found the old priest awaiting us for some presents in return for the old images he had allowed us to take from outside the morai. We gave him several articles of clothing with which he was more pleased than if we had given him money.

Visits Kapuhiolono and Erects Monument to Captain Cook's Memory

July 17. Sunday. Fine. Went to see the monument erected yesterday to the memory of Capt. Cook by his Lordship (Byron) on the top of the black lava hill above Karona, where it is said his body was divided amongst the chiefs. This monument is simply a capstan bar from the ship, painted white and fixed on a heap of lava that had formerly been raised by the natives when they made the square wall round this spot, which they now for years look upon as consecrated ground.

This bar stands about sixteen feet high, with a small copper plate at the top, having the following inscription upon it:

In memory of Captain James Cook, R.N.,
who discovered these islands
in the year of Our Lord 1778
This humble monument is erected by his fellow countrymen
in the year of Our Lord 1778 [EDITOR'S NOTE: should be 1825]

THE *BLONDE* SAILS FOR TAHITI

July 18. At 12:10 a.m. weighed anchor and made sail with light breezes for Otahite, leaving these interesting islands after a short stay of little more than two months. With the exception of a few chiefs of high rank, who have adopted a little of the European style of costume, the natives still retain their manners and customs nearly the same as related by the unfortunate Cook who fell a sacrifice to their barbarity. The dagger which is said to have committed the fatal deed, I am now informed, is in the possession of a gentleman on board, who is also informed of every particular circumstance relating to the misunderstanding which took place between the parties at the time. But I am inclined to doubt if they had such a thing as a dagger at that period, or why it should be held secret from Vancouver and other navigators who wore equally anxious to gain all information respecting the occurrence and with what weapon the fatal deed was committed. The small English drum presented by Cook to the King is now in his Lordship's possession, with some other things which have been in the morai for many years and were given to Lord Byron when he visited the morai a few days ago.

(Here follows in Macrae's manuscript a list of 25 fruits or plants, indigenous or introduced from abroad, which is hardly worth reproducing. He states that halla (hala) pineapple is "not plentiful." What would Macrae say if he returned to the islands in 1922? He would find square miles of pineapples.)

July 19-23. Fine fresh breezes from E.N.E.

July 27. Fine. Breezes E. by S. Crossed the Equator.

DISCOVERY OF MALDEN ISLAND

July 30. Sighted land, at 12 noon. Lowered two boats and landed through the surf on a fine sandy coral beach covered with shells, most of which were the cockle kind of very large size weighing over 15 lbs. We separated to explore the island, and found it low and flat, covered with vegetation of no great variety, being mostly sidas and a few bushes of tournefortia. The island was covered with birds, remarkably tame and easily taken in our hands.

We saw remains of huts built of coral slabs without any timber, and we gathered therefrom that the island had been inhabited, probably by ship-wrecked people and never discovered. At one of these huts, the surveyor left a sealed bottle which he had brought for the purpose, containing some written inscription, what, I know not. I saw no signs of cultivation. There was a little fresh water in holes among the coral rocks, a good distance from the sea. Just when our stay was becoming most interesting, the recall gun from the ship was fired, so we could not find out any more about the island. We thought that we might perhaps find something concerning the unfortunate Mr. de la Perouse and his crew, about whom the world still remains doubtful as to what part of these seas they were lost.

This little island, which is a discovery of the *Blonde*, and since named Malden Island, in honour of our surveyor, is about 12 miles in circumference, is very low and scarcely seen 15 miles distant out at sea. It abounds with sea birds, but without quadrupeds except a small brown shiny kind of short-tailed rat, very wild, that ran into their holes at our approach.

HERE ENDS JAMES MACRAE'S DIARY

CONDENSED LOG OF H.M.S. *BLONDE* AFTER LEAVING MALDEN ISLAND

1825		
August 1	—	Passed Starbuck Island.
August 8	—	Reached and landed at Mauke, one of the Cook group.
September 4	—	Passed Juan Fernandez.
September 6	—	Reached Valparaiso.
September 29	—	Anchored at Talcahuaria (Concepcion).
October 12	—	Sailed from Concepcion and reached Valparaiso 13th.
December 5	—	At Coquimbo.
December 29	—	Doubled Cape Horn.
1826		
January 23	—	Arrived at St. Helena and stayed there until 28th January.
March 7	—	Rescued the survivors of the wrecked ship *Frances Mary*, bound from New Brunswick for Liverpool.
March 15	—	Anchored at Spithead after an absence of seven months and fifteen days.

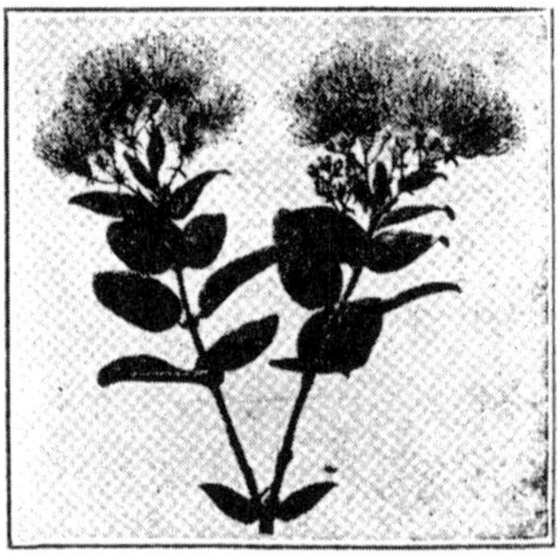

CONCLUSION

William Wilson's *With Lord Byron at the Sandwich Islands* might be one of the first books published in Hawai'i based directly on a manuscript located in Europe. The process through which the manuscript was used to create the extract is not recounted in the book. Wilson was a British merchant living in Hawai'i, on Prospect Street in Honolulu to be precise, and his connections to trade and to Britain would have made it easier for him to either travel to Britain or to find someone in Britain to visit the Royal Horticultural Society's archive. Wilson's extract was created during the Territorial period in Hawaiian history, following the overthrow of the Hawaiian monarchy by primarily American business and military interests. Wallace Farrington was the governor, having been appointed by President Harding. The domination of the Hawaiian economy and culture by plantations, tourism, and the military was well established. In this context, Wilson's publication arose within a cluster of people and organizations in Hawai'i connected to Hawaiian historiography, or perhaps more accurately to antiquarian-inspired activity. These were the people who would have encouraged his projects, helped with his research, and read his results.

Thomas Thrum, an Australian and founding member of the Hawaiian Historical Society, began publishing *Hawaiian Almanac and Annual* in 1875 and *Paradise of the Pacific* in 1888. Both publications included short accounts of historical events, people, artifacts, and anything else related to Hawaiian history. The goal was to collect fragments and manuscript material that would not otherwise be available to the interested reader. In the 1922 edition of the *Hawaiian Almanac and Annual*, Thrum notes:

> Wm. F. Wilson presents a valuable contribution to Hawaii's early history in his pamphlet entitled: "With Lord Byron at the Sandwich Islands in 1825," being extracts from the diary of James Macrae, Scottish Botanist,

unearthed from the archives of the Royal Horticultural Society of London. Printed in Honolulu for the author, 75 pages, ill. 8 vo., paper. (Thrum 83)

Such notices of relevant publications were commonplace; the *Almanac* functioned as a general portal for its members to materials relevant to Hawaiian history. The publication was a shared space that supported a community of shared interests.

In addition to local publications, extensive publishing activity also occurred off-island. In 1888, New York printing house C.L. Webster published King Kalakaua's *The Legends and Myths of Hawaii*. By the 1910s popular books on Hawaiian myths and legends had been published, such as Thomas Thrum's *Hawaiian Folk Tales* (1907), first published in Chicago, and W.D. Westervelt's *Legends of Gods and Ghosts* (1915) and his *Hawaiian Legends of Volcanoes* (1916), which were both originally published in Boston. A chair in the study of myth had been created at Vassar College in 1920. The chair was first held by Martha Beckwith, who became a significant scholar of Hawaiian legends and published her *Hawaiian Mythology* in 1940.

Honolulu's Bishop Museum was founded in 1889 and their press was created in 1892, with its first publication being a preliminary catalog of their collection of "Kahilis. Feather Ornaments. Mats and Kapas." In 1899, they published William Tufts Bingham's *Hawaiian Feather Work*. Between 1916 and 1920, Bishop Museum Press published Thomas Thrum's compilation and translation of Abraham Fornander's newspaper writings, entitled *Collection of Hawaiian Antiquities and Folk-lore: the Hawaiian Account of the Formation of Their Islands and Origin of Their Race, with the Traditions of Their Migrations, etc., as Gathered from Original Sources*. By the early 1920s there existed a substantial printed collection of works connected to Hawaiian history and culture.

Finally, the Hawaiian Historical Society was founded in 1892, and their annual reports began the following year. The reports included society business, a list of active members, and reprints of papers on topics of interest to the group. The Society brought together some of the influential historians of the time, including W.D. Alexander, Thomas Thrum, W.D. Westervelt, and Nathaniel Emerson. Wilson joined the Hawaiian Historical Society in 1897, and while he did not consistently appear on the member list, he remained connected to the group.

The 1922 *With Lord Byron in the Sandwich Islands*, then, exists near the tail end of this period of publication. Wilson had already published *David*

Douglas, Botanist at Hawaii in 1919 and would publish *Reminiscences of John Cook, Kamaaina and Fortyniner* in 1927. Wilson's recovery of Macrae's journal remained the only version of that journal to be printed and, more broadly, remained one of only a very small number of books to be written about the voyage of the *Blonde* to the Hawaiian Islands.

But Wilson's recovery of Macrae's journal was not perfect. Wilson's extract was limited and changed the content and tone of Macrae's original journal in significant ways. A complete transcription of the original journal is needed to recapture what Macrae wrote during those two and a half months in 1825 while the *Blonde* was in the Hawaiian Islands. Wilson's extract is commendable and is a worthwhile historical document from the 1920s. However, access to the original journal, as presented in this publication, will expand the understanding of European voyages to the Hawaiian Islands in the 1820s.

Changes

When approaching a book published in the 1920s that was based on a manuscript written a hundred years earlier, it is worthwhile to consider how the printed text compares to the original. Are the same events included, and are they included in the same way? Are different people or groups made to look better or worse? How does the passage of time and presentation of history influence what information should be included? As suggested by the differing accounts of the royal funeral included in the Introduction, Wilson's extract is not a direct transcription of Macrae's journal. Wilson does not explain why he included or excluded certain parts of the journal. His editorial changes are not surprising, though, especially if the goal is to increase readability for a general audience. Macrae's writing is often long-winded, he uses non-standard abbreviations, and his spelling and grammar are not typical for modern English. The journal was not written to be published later; instead, it was a way for Macrae to report to his employers at the Royal Horticultural Society.

Wilson also did not publish the journal in its entirety. There are roughly 27,000 words in the extract compared to roughly 43,000 in that section of the journal. That is to say, Wilson's printed book has about 60 percent of the original text. Almost all of the sentences have been rewritten, details have been left out, and the editorial choices have affected the tone and content of Macrae's account of the voyage. In addition, the time spent in the Hawaiian Islands represents only half of Macrae's entire manuscript journal. Wilson produced a heavily edited extract, not a transcript. The changes are worth discussing.

Errors

Some of the changes in Wilson's extract are simple typographical and transcription mistakes. For instance, Wilson's version has Mr. Young, a European who advised Kamehemeha I, on the island twenty-six years before the *Blonde* arrived, when in fact Macrae writes that it was thirty-six. In another passage, the extract refers to "the captain of an American whaler," whereas the journal refers to "the Captain of *American Whales*," (page 38) referring to the name of the ship. Likewise, in the June 15 entry, after his group was caught in the surf while trying to land, Macrae writes in the journal, "When we had partly shifted ourselves with dry clothes, & finished rubbing our forearms free from the salt water, we examined the rest of our luggage to ascertain if anything had been washed away by the surf while in our confusion landing" (page 85). The extract, in contrast, simply states, "We got into dry clothes as far as possible and dried our firearms" (page 189). The printed passage is much shorter, but also mistakenly uses "firearms" instead of "forearms," which is significantly different, and while firearms makes some sense in the shorter version, it makes no sense in the original text. Transcription errors, however, were not common in Wilson's version, as most of the differences between the original journal and Wilson's publication were intentional changes to the text.

Rewriting

A much more common type of change from the manuscript to print were changes in terms and phrases. Throughout the extract, Wilson shortens Macrae's long and sometimes convoluted sentences and removes complicated phrases. Examples can be found on every page: "Be this as it may" (page 49) is changed to "anyway" (page 158), while "at a little distance from our hut" (page 76) becomes "near here" (page 182). Likewise, when recounting the battle of Nuʻuanu, Macrae's introductory phrase, "The circumstance related to us was nearly as follows," (page 73) becomes "It happened as follows" (page 180). On May 27, "Being informed during the day..." (page 70) becomes "Hearing..." (page 175). Changes of this type occur throughout the text.

Sometimes the vocabulary is changed. For instance, "the wood" is typically changed to "the forest," although in one instance the "wood which we had come through" becomes the "road" in the extract. Likewise, while Macrae writes of how a volcano is "now closed" (page 65), Wilson updates the term so that it is "now extinct" (page 172). The claim that there are hummingbirds in

Hawai'i, on the other hand, is corrected with a footnote (page 163), allowing the error to exist in the extract but without misleading the reader.

Wilson's text uses diverse editorial strategies with terms and names. He updates much of Macrae's spelling and terminology, changing "tarro ponds" to "taro patches," for instance. With proper names, however, Wilson typically maintains Macrae's spelling while adding the updated name in the section headers. He leaves "sanders wood" unchanged in the text but changes it to "Sandal Wood" in the section header. "Tantalus" is used in one section header, although it is unnamed in the journal. Other place names are mixed, such as using "Diamond Head" as well as Macrae's "Diamond Hill." Likewise, Wilson's extract uses "Sandwich Islanders" instead of Macrae's original terms "Sandwich Island people" or "Sandwich people." The result is a more consistent, tight writing style, but also one that moves away from Macrae's original.

Reduction of Detail

Tightening Macrae's writing results in a noticeable reduction in the number of words used to describe people, places, things, and events. For example, the journal refers to Isaac Davis's "wife & three daughters" (page 71), while the extract refers to his "wife and daughters" (page 175). Likewise, "nearly dusk" (page 113) becomes "dusk" (page 208), and "the frightened natives" (page 100) becomes "the natives" (page 201). Wilson also removes most of the information about weather and temperature. Every day while at the islands, and almost every day throughout the voyage, Macrae notes the temperature in the morning, at noon, and in the evening, and describes the weather conditions in some detail for most days, sometimes including changes during the day. The information can be considered uninteresting, but it is a persistent part of Macrae's journal and helps to ground the journal in the material conditions of the ship and the islands.

Wilson likewise removes many of the references to Macrae's botanical collection. Macrae constantly writes about his plants, seeds, and specimens, his work shifting them to dry paper, his need to find space on shore to work on new additions, his efforts to keep the gift plants alive, and so on. Fragments of these activities remain in the extract, but most of his botanical activities have been left out. A typical example is when Macrae describes how "[I] laid my specimens of plants brought home in paper, & in the afternoon sorted & packed up my specimens of lava, as also arranged & shifted others gathered before I left" (page 112). None of this text occurs in the extract. One implication of Wilson's decision is that Macrae becomes less of a working botanist to the readers of his extract. The phrasing is shorter, the statements are clearer, but the text loses character. He does not spend much time collecting, and the details from when he actually collects are minimized.

The loss of detail connects to a general decrease in the tentativeness found in Macrae's journal. Instead of being "creditably informed" of how the gambling game *Nooah* works (page 98), in the extract he is simply "told" (page 200). Likewise, the journal describes coconut trees, "some of which are very great height & show the appearance of old age" (page 82). In the extract, they are simply "of great height and age" (page 187). This type of change is significant insofar as Macrae, as a product of the Scottish Enlightenment, is much more likely to be skeptical of absolute claims and more likely to recognize that sources need to be named and their reliability needs to be assessed.

In the printed extract, details of people are frequently lost, which results in blander, simpler accounts of those whom Macrae comes into contact with. Examples can be found throughout the document: "this Otahitian" (page 56) changed to "this man" (page 164), "a few wretched ragged foreigners" (page 62) becomes "a few ragged foreigners"(page 170), and Macrae going on shore "with my friend Mr. Forder" (page 60) becomes simply "with Mr. Forder" (page 168). The extract does not mention many personal descriptions of people, including one where the hair of the Hawaiian Queens was described as "straight & of a brownish black colour, neatly put up after the manner of the Europeans" (page 43). In most cases, these changes save small amounts of space on the page. However, they also create a general flattening of the people in the narrative.

Some people are excluded completely; some people have their roles diminished. Mr. Charlton, the recently-arrived British consul, has a similar status in the journal and the extract. However, Mrs. Charlton and her sister, who are mentioned three times in the journal, are completely gone in the extract, except for one reference to Charlton's "family" (page 182). The single reference in the journal to Mr. Sabine (page 77), the Secretary at the Royal Horticultural Society and Macrae's supervisor, has been removed, along with the fact that Macrae describes how he spends "all day writing out my accounts, as well as letters for" him. Much of the complexity of the original journal is lost.

Beyond a general desire to tighten the writing and increase clarity, there are sometimes other reasons why Wilson may have wanted to shorten the text. For example, "the young King" (page 42) becomes "the king" (page 152), while "the young Princess his sister" (page 76) becomes simply "the young Princess" (page 182). That last change could have political implications given that their relationship was highly controversial, especially to the missionaries who opposed their plans to marry. This larger issue is not mentioned by Wilson.

Surprisingly, some subjects touched on by Macrae are reinterpreted by Wilson in a manner entirely in opposition to Macrae's intent. Take alcohol, for example, where Wilson typically replaces Macrae's positive evaluation of alcohol with a negative one. On one occasion, Macrae describes how the captain of a whaler that had wrecked on a reef had "enjoyed himself too freely," an episode that Wilson recounts as well, but he excludes the second half of the description: that the captain had been drinking "with his friends before going on board." Another example is when Macrae is part of a larger group on an excursion to what is now Ford Island. The party becomes separated, they have a difficult time landing, and by the end of the day they are quite tired and frustrated. Wilson includes Macrae's description that "after having sat down to dinner on the mats spread out on the grass, matters were soon again restored to perfect harmony," but fails to finish the passage where "the wine bottle passed round freely to every one of the company to drink each other's health without form or any kind of ceremony in respect to persons." (page 61). At another time, when Macrae is hiking with a group, he notes, "our own stock of provisions being now reduced to only three biscuits & about a pint of brandy left in Mr. Talbot's flask, which we found very acceptable when taking a mouthful to warm us, while we had wet clothes on" (page 99). Wilson's extract omits Macrae's description of the social and medicinal uses of alcohol. These changes could be explained by the historical context of Wilson's extract, which was published in 1922, three years after the 21st Amendment made drinking illegal in the United States. Wilson describes in a footnote an episode in Macrae's journal where marines from the *Blonde* had "got pretty well loaded up with okolehao (native gin)" and had been punished (page 185). Wilson recounts a recent event where 52 sailors were incarcerated in downtown Honolulu following a night of drinking. He remarks, somewhat sarcastically, how the episode "happened in 'dry' America."

A more extreme example of editorial simplification is when Macrae describes the preparation and eating of certain foods that he finds disgusting. The first instance concerns the preparation of a dog. The choice of food is not completely removed, but it is sanitized, perhaps to make the text more palatable to Wilson's anticipated readers or to remove Macrae's harsh reaction to the meal and subsequent judgment of Hawaiian culture.

Wilson's Extract, page 214	**Macrae's Journal, pages 121–122**
We saw a dog being cooked for one of the chiefs, who consider them better eating than any other animal.	During the time we remained on shore we had the opportunity of seeing a dog getting cooked for one of the chiefs who consider them better eating than any other of their animals. Which is done in the same manner with cooking hogs. The throat is first cut & when the bleeding ceases they then place the animal over a large fire to get the body deprived of the hair, instead of using hot water, & when the inside is cleaned, it is afterwards filled with hot stones & buried in the ground for some length of time covered over with leaves & earth. Their filthy manner of killing dogs, which is the employ of severals at the time & are immediately after cooked & eat by the chiefs, is too disgusting to describe, particularly the ravenous way they consume the entrails, while the rest is cooking.

In another example, Macrae describes native Hawaiians eating raw fish procured while they were on the *Blonde* off of Maui.

Wilson's Extract, pages 185–186	**Macrae's Journal, page 80**
June 9. Being near the island of Mowee, a boat was sent on shore with Joe Banks, to procure fish. He returned with a quantity, fresh and salted, which the natives immediately began to devour raw, with some poi and taro, **not even wasting their gills or entrails.** These parts are considered by them to be the most delicate parts of the fish, and were what was generally eaten by the queens and chiefs. Among others doing this was Manaware, whom we had brought out from England. When I, disgusted,	9. A fine morning with fresh breezes from W. by South. At 4pm the wind ceased to a calm, being now near in shore off the land of Mowee, a boat with Joe Banks was sent on shore to some huts near the sea beach to procure fish, who after returned at five o'clock with a large quantity of fresh & salted, which the natives immediately after getting them on board, began to devour the fresh in a ravenous manner raw, with some poi & tarro, **not even wasting their gills or entrails,** but considered

expostulated and told him he ought to have shown his countrymen how he had seen us eat our meals, he replied, somewhat offended, that he and his countrymen liked fish in that way best, adding that "He saw plenty of poor people in England, but we see none here; that they got plenty of poi, taro and fish and no want for anything like many a man at home."	them the most delicate parts of the fish, & what was generally eat by the Queens and chiefs. Among those eating fish in this disgusting manner, we noticed to our surprise Manaware who we had brought out from England seated in the midst of them with his shirt sleeves rolled up to his shoulders, eating & tearing to pieces with his teeth the gills & entrails having his mouth, hands, & fingers besmeared with blood, from his not using a knife & fork a thing yet never adopted by the Sandwich islanders I was so much disgusted on seeing this fellow, who of late had been accustomed of seeing a different way of living, that I could not help be without telling him that he ought to have shown his countrymen how he had seen us eat our meals, to which he replied somewhat offended, that he & all the people of his country liked fish in this way best, observing at the same time, he see plenty of poor people in England, but we see none here that they got plenty of poi, tarro, & fish, & no want for any, like many poor man at home.

In both encounters, Macrae provides more details and harsher judgments than Wilson's extract would suggest. The basic event still exists, but many of the details are muted or gone. Overall, the text, including both the Hawaiians and Macrae, becomes less intense.

On Diamond Hill

As a final example of Wilson simplifying Macrae's journal, consider these two descriptions of Macrae returning home from Diamond Hill with a lad from the *Blonde* named Mantle and a Hawaiian guide. As they descend, Macrae's hat is blown off and when Mantle goes to retrieve it, he mistakenly picks up a skull instead of the hat and is horrified to encounter piles of human bones.

Wilson's Extract, pages 172–173	**Macrae's Journal, pages 65–66**
On our way home, in a high wind, my hat was blown off, and Mantle went down the cliff to retrieve it. It was nearly dark and he called out to myself and the native guide not to leave him behind. Afterwards he told me that in picking up the hat, he had picked up a large human skull. At first he had not noticed what he had picked up till he saw the hollow part of the eyes, and had called out in fright and thrown the skull away. On my expostulating he said he could get plenty more, as there were a lot there. This was explained to me by **the American I had hired as cook and who had been on the island for six years,** by his telling me that in former times, the chiefs used to take their criminals upon the top of Diamond Hill to put them to death by throwing them over the precipice, where they were left unburied in the hollow, and thus Mantle had met with many skulls.	While returning home along the narrow ridge on top of Diamond Hill, where it is very abrupt & rather difficult to pass. The wind at the time blow rather strong which while picking up some specimens of stones carried away my hat over the outside declivity towards the sea, when Mantle immediately went after it, at the nearest place he could descent of his own accord contrary to what I wished, as it was now getting dark & the native with us would bring it me in the morning. The native who remained with me while I was getting down to the bottom had scarcely ever uttered a word since Mantle had been gone, till after waiting for some time on getting to the foot of the hill, Mantle in the dark began to call out aloud for us not to leave him behind & running towards us much frightened, which now the native perceived & laughed heartilty holding out his hand in the direction where Mantle was coming from, using the words "houry maiti" (no good). When Mantle came he told me the cause of his fright, was the discovery of having picked up a large human skull in the place where he found the hat, but being in a hurry & dark he did not perceive what he had got till thinking he was near me, when he then looked at it & saw the hollow parts of the eyed, which made him call out to us as he did & then unfortunately threw it away. But on enquiring whether he thought he could find it for me again, he said it made no difference as the same place where he found it was full of others, which an account of our having a native with us, we durst not venture on taking any away. At 8 o'clock

	when we got home I related to Mr. Forder Mantle's fright in going in search of my hat which afforded some mirth at his expense. I learned from **an American who had been on the island for six years & now with me since I came on I had hired to cook** & make himself otherwise useful that the chiefs used in former times to take their criminals upon the top of Diamond Hill, to put them to death, by throwing them over the precipice when they were left above ground without being buried in the ravine that Mantle met with so many skulls. Thermometer at 6am 73 at 12 86 & at 7pm 76.

The extract is barely a third of the length of Macrae's original journal entry (425 words down to 177) and most of the details are lost. The native who had accompanied Macrae and Mantle is less revelant, which means that Macrae's attempt at writing down the Hawaiian phrase is also gone. Wilson's passage leaves out the suggestion, which some readers might find troubling, that Macrae wanted to find more bones to take back with him, although he admits that he would not have done so anyway because the native guide was there. Finally, the extract misses the humor that Macrae and Mr. Forder found in the episode afterwards. Instead, the reader is offered a rather mundane episode with some interesting details about how Diamond Hill was used by the native Hawaiians.

The trip to Diamond Hill is an example of a tendency by Wilson to leave out or reduce Macrae's interactions with other people. Macrae's conversations with Lord Byron are often not included, such as when he informs Byron of his plans for the day, or when one describes to the other what happened during a trip on shore. Few of these episodes are compelling reading, but the reader also loses the sense of Macrae's day-to-day life on the islands. He complains, he hears stories about the islands or about what other people have done, he arranges passage to and from the shore, and a host of other activities that are noted in the journal but left out of the extract.

Francisco de Paula Marín

Another substantial editorial change occurred with Francisco de Paula Marín, and Macrae's conflicts with him. Marín's place in Macrae's journal is downplayed by Wilson. The extract does not include Macrae's comment that Marín "acted as our interpreter" (page 45), or that the navigator Otto von Kotzebue had talked about Marín. Wilson does include the conflict between Marín and Macrae over the plants that Macrae had brought with him. Even here, however, much of Macrae's criticisms are toned down in the extract.

Wilson's Extract, page 181	**Macrae's Journal, pages 74–75**
Marin, the Spaniard, previously mentioned, has had a grape vine since 1814, which he brought from California, where he had been with Adams as linguist, and had planted in his own ground near Hanarura, and got the king to perform the charm of taboo to prevent any being stolen. This taboo still continues in force to the present time, but although he has cultivated the vine with success for the last two years, and made more than two casks of wine from the fruit, sooner than give the slips to his neighbors, he has burnt the prunings every year. But Mr. Charlton has given to different individuals, several fine grape vines which he brought with him from the port of Valparaiso, and also an assortment of vegetable seeds, the gift of the Horticultural Society, London.	The Spaniard (Marin) whom I have already had occasion to mention, has had the grape vine in his possession since 1814 from California at which place he had been with Adams as linguist in one of the Sandwich Islands vessels to bring over cattle, horses, & mules &c., which on his return he planted the grape vine in his own ground near the town of Hana-rura & got the King to perform the charm of taboo to prevent any being stolen which till the present time still continues to be in force. So selfish is the disposition of this man that although he has succeeded to cultivate the grape vine with much success & within the last two years to make more than three casks of wine from the fruit, that sooner than he would part with slips to his neighbours he actually burned the prunings every year rather than oblige those who wished for to cultivate them, that he might enjoy the monopoly arising from them himself, which will now soon be out of his power from the liberality of Mr. Charlton who has given away to different individuals several fine grape plants brought with him from the Port of Valparaiso, besides an assortment of vegetable seeds from England presented to him by the Horticultural Society

Marín is not described positively in either version; in the journal, however, Macrae is much more explicitly critical of "the Spaniard's" selfishness. Other editorial decisions can be noted in this passage as well. The fact that Marín and Adams had sailed to California on a Sandwich Island vessel is significant but missing.

It should also be noted in the passage above that Wilson excludes mention of the Hawaiian sailing ships on other occasions as well, which also simplifies both the narrative and also the changes that have occurred on the islands. In Wilson's version, the native Hawaiians tend to be less Westernized and less likely to understand or use Western technology.

Macrae's conflict with Marín is one of several he has throughout the voyage that are missing or downplayed in the extract. He is often upset by promised resources, canoes, huts, and time that fail to be provided. He complains about native guides not appearing on time and the gentlemen aboard the *Blonde* treating him poorly. The result is that Macrae in the extract comes across as a more pleasant person than is perhaps warranted. The reader is unaware of the tensions on board the ship, or Macrae's awkward status as a supernumerary aboard a British man-o'-war. One example of many where Macrae is frustrated is when he tries to take several Hawaiian artifacts from a cave near Kealakekua Bay.

Wilson's Extract, page 213	**Macrae's Journal, page 119**
While hunting in the corner for curiosities, we found an old fishing rod (which was given to me) and a club. The rod was chiefly made of human bone, and the club made of wood shaped like a thigh bone. However, Lord Byron seeing the chief at his hut, prevented me from keeping the rod, as we had no permission to take anything away, so to our disappointment, it had to be replaced where we found it.	While searching the corners for curiosities two of the party found an old fishing rod & a club the former chiefly made of human bone the other of wood in shape of the thigh bone. The fishing rod was given to me, as a difficulty appeared to venture in taking it away without asking the chief who had power to give it, which trouble I willingly took upon me by getting it for myself, but was on seeing the chief at this hut, prevented from having it by the interference of His Lordship, who unluckily for me happened to be at the time present & who requested the club as well as the fishing rod to be restored to go where it was taken from. I felt as may be imagined not a little

	disappointed for his interference in so trifling a thing, that I was sure if the chaplain or any other of the sun worn officers who since we had been at the islands were considered from their rank on board to be chiefs of some consequence like themselves, had wished for it they would not have received no denial.

Soon after, Macrae obtains some wooden statues from a nearby *morai* [*heiau*] and is later visited by the priest asking for payment for the items.

Wilson's Extract, page 215	**Macrae's Journal, page 124**
On board we found the old priest awaiting us for some presents in return for the old images he had allowed us to take from outside the morai. We gave him several articles of clothing with which he was more pleased than if we had given him money.	At four in the afternoon returned on board when we found the old priest waiting for us to make him some present for the images he had given us at the morai, for which we gave him a few articles of clothing & seemed perfectly satisfied more so than if we had given him money. Having been but a short time on board when I was sent for by the commanding gentleman at the time, for me to take away my wooden gods, who he said were only for the cook to boil the coppers in the galley. Forgetting himself that he was more lucky on visiting the morai before me to get feather images of more consequence besides other curiosities that would be denied me like the fishing rod, because I was not considered by the natives to be a chief like himself & others on board.

The interaction with the priest is largely unchanged in the extract, if a little shorter. However, Wilson excludes Macrae's interaction with the commanding gentleman immediately afterwards, who makes fun of Macrae's artifacts. Macrae writes of how lucky the gentleman was to get better artifacts because he had visited the *morai* first and how unlucky Macrae was because he was not considered to have the status equivalent to a chief. Macrae's resentment towards the gentlemen on the *Blonde* is an ongoing theme in the journal, and one that Wilson generally does not include in the extract. These editorial decisions make Macrae seem less whiny and covetous. As with the other editorial changes, however, Macrae does not become more pleasant so much as he becomes less of a character in general.

Narrative Scope

The final and greatest difference between the journal and the extract is the scope of the narratives themselves. The original journal exists in a much broader field: namely, the entire voyage from London to Brazil to Chile, to the Galapagos Islands, to Hawai'i, and then back to Chile, St. Helena, and Britain. The visit to the Hawaiian Islands extracted by Wilson was in fact part of a larger voyage that had grander strategic goals. His choice to omit the other parts of the voyage means that the reader is deprived of this important context. HMS *Blonde* was a frigate and was in the Pacific as part of a general show of force in South America. Byron's diplomatic mission to the islands was ancillary to this. In fact, the time at the Hawaiian Islands is roughly half of Macrae's journal.

By offering a slice from Macrae's complete journal, Wilson, and by implication the new transcript offered in the current publication, downplays the connections between the Hawaiian Islands and the rest of the world.

While a complete transcription of Macrae's journal would be useful to place Macrae's time in the Hawaiian Islands in a larger context, it is also possible to use other publications to the same purpose. Publications describing 19th-century European activity in the Pacific include similar topics to those in Macrae's journal. Whaling and trading ships were common and explorers continued to arrive at the islands. Similar types of publications describing 19th century European exploration of the Pacific recount some of the same events as Macrae recorded in his journal, but with a larger historical context.

Otto von Kotzebue visited the Hawaiian Islands in 1816–'17 and again in 1824–'25. In his second voyage, Kotzebue arrives in early November of 1825, six months after the *Blonde* had left. Kotzebue includes a brief description of the funeral, a few details of Byron's visit, and some account of how the missionary "Bengham" has gained a dangerous influence over the new king and his regent (Kotzebue and Eschscholtz 254–55, Volume 2). Having also been to the Hawaiian Islands in the 1810s, Kotzebue is able to reflect on how the islands have changed, although his work does not have the level of detail typical for published exploration books of the time. Captain Frederick Beechey would arrive in the islands on the *Blossom* the following year, in May 1826, and provide further updates, referencing previous published accounts as well as offering additional first-hand descriptions. But Beechey's voyage also places the Hawaiian Islands in the context of Pacific and global structures. His voyage is focused on finding the Northwest Passage, and the printed account includes the entire voyage from Britain to Brazil, Easter Island, Tahiti, Hawai'i, the Northwest coast of North America, and then back to Britain. Like Macrae, Kotzebue and Beechey were part of global voyages, which is relevant for understanding the time that each of them spent in the Hawaiian Islands.

Wilson's extract tends to downplay the net of interactions that the islands were already a part of, and it obscures the persistent impact of foreigners and the ongoing connections that existed between the islands and other places. By the 1820s, there were many ships connecting the islands to the Americas and Europe in reliable ways that allowed information, materials, and people to move. People living on the islands had heard about the death of their King and Queen from a whaling ship almost two months before the *Blonde* arrived. People on the *Blonde* were also aware that Mr. Charlton, the new British Consul, was making his way to the islands, and in fact arrived from Valparaíso on the *Active* the week before the *Blonde*. Native Hawaiians were living in the Pacific Northwest as employees of the Hudson's Bay Company, and many others were working on whaling and trading ships. Macrae writes of sending seeds back to London via Mexico, and his journal also describes going on shore to "write dispatches for His Lordship to go home by the *Active* bound to Valparaiso in a few days" (page 75). Europeans and Americans also played roles in the local infrastructure. For instance, Macrae notes that Novara "is besides an innkeeper an auctioneer & commission merchant, which brings him the customs of all masters of vessels coming to this part..." (page 48). Wilson merely refers to Novara as the owner of the inn. The extract likewise makes no mention of Lord Byron endorsing a bill for Macrae for £20 (page 77), which suggests kinds of financial interactions that were possible in Honolulu at the time. Contrary to what may be inferred from Wilson's editorial changes, there was a global infrastructure of which the Hawaiian Islands were a part. Of course, Macrae was not aware of everything that was happening—it is doubtful, for instance, that he knew much about native Hawaiians working in North America—but his journal nonetheless provides evidence of some of the connections that existed at that time.

Conclusion

For people living in the 19th century, and the reading public in particular, James Macrae barely existed. After the *Blonde* returned to Britain, his journal was presented to the officers of the Royal Horticultural Society and duly placed in its archive. It never became part of the public record and was not consulted when the official account of the *Blonde's* voyage was created. Macrae was sent to Ceylon to work at the Colonial Gardens and would die there a few years later. In print, he existed as part of longer list of names of the people on the voyage.

It is only with Wilson's archival recovery that Macrae has become known. But Macrae is different in the 1922 version. He is more of an observer without any real connections to the land or the sea. His interactions with Byron are minimized. Macrae is still disgusted with certain foods eaten by the Hawaiians and still upset at the behavior of Marín and the attitudes of the gentlemen aboard the *Blonde*, but in Wilson's extract the reactions are muted. Macrae is no longer primarily a botanist working for a royal institution, spending much of his time focused on his assignments. Rather, he is a disembodied voice describing the world, but not really participating in the life of the people who are encountered or the land that is walked upon. The physicality is gone. The people, both Hawaiian and European, occur less frequently, and when they do occur, they are more superficial. The most substantial change from the journal to the extract is a general bleeding away of descriptions. We lose much of Macrae and the complex history he was part of. But Wilson still contributed something, he made Macrae's journal part of a larger set of texts connected to the voyage of the *Blonde*. These narratives, which include Callcott and Stewart, discussed in the Introduction, help to make the *Blonde's* visit to the islands more complex and point out variations in how the visit was described and evaluated. Those texts, in turn, are tied to particular perspectives—missionaries, naval officers, botanists, travel writers, antiquarians—in a print world dominated by European and American writers. Hawaiian voices also exist, such as Samuel Kamakau, John Papa ʻĪʻī, and those writing in Hawaiian-language newspapers, the first of which began publishing nine years after the *Blonde* left the islands.

The transcription of the original journal printed in this publication is a further addition that is closer to Macrae's original account. But Macrae's is likewise an incomplete picture, conveyed from the viewpoint of a Scottish botanist of little status who was primarily concerned with producing an account for his distant employer. He had little interaction with Native Hawaiians and a limited understanding of what was happening either in the Hawaiian Islands or elsewhere in the world. Nonetheless, accounts such as Macrae's can enhance our conversations concerning how to understand the history of the Hawaiian Islands by providing alternative descriptions and personal observations. There is much more to be done.

Bibliography

Barnes, James J., and Patience P. Barnes. "Reassessing the Reputation of Thomas Tegg, London Publisher, 1776–1846." *Book History* 3 (2000): 45–60.

Beasley, H. "A Carved Wooden Figure from Hawaii." *Man* 32 (1932): 33–33. *JSTOR*, doi:10.2307/2789605.

Bingham, Hiram. *A Residence of Twenty-One Years in the Sandwich Islands; or, The Civil, Religious, and Political History of Those Islands: Comprising a Particular View of the Missionary Operations Connected with the Introduction and Progress of Christianity and Civilization among the Hawaiian People*. Hartford: Hezekiah Huntington, 1847.

Bloxam, Andrew. *Diary of Andrew Bloxam, Naturalist of the "Blonde" on Her Trip from England to the Hawaiian Islands, 1824–25*. Bishop Museum Press, 1925.

Callcott, Maria. *Voyage of H. M. S.* Blonde *to the Sandwich Islands, in the Years 1824–1825. Captain the Right Hon. Lord Byron, commander.* London: John Murray, 1826.

Cutter, Donald. "The Spanish in Hawaii: Gaytan to Marin." *Hawaiian Historical Society*, vol. 14, 1980, 16–25.

Dampier, Robert, and Pauline King Joerger. *To the Sandwich Islands on H.M.S.* Blonde. University Press of Hawaii for Friends of the Library of Hawaii, 1971.

Elliott, Brent. *The Royal Horticultural Society: A History 1804–2004*. Chichester: Phillimore, 2004.

Gast, R. H. *Don Francisco de Paula Marin: The Letters and Journal of Francisco de Paula Marin*. (A. C. Conrad, Ed.) (First edition). Honolulu: Hawaiian Historical Society, 1973.

———. *Don Francisco de Paula Marin: a biography*. Honolulu: University Press of Hawaii for the Hawaiian Historical Society, 2002.

Hamilton, Claud. "Two Early Monuments to Captain Cook." *The Geographical Journal* 57.1, 1921, 34–36. *JSTOR*, doi:10.2307/1781203.

Historical Commission of the Territory of Hawaii, and Hawaii. "Secret Instructions to Lord Byron, Sept. 14, 1824." *Report of the Historical Commission of the Territory of Hawaii for the Period Ending December 31, 1926*. Honolulu: Historical Commission of the Territory of Hawaii, 1927. 19–20.

Huish, Robert. *A Narrative of the Voyages and Travels of Captain Beechey to the Pacific and Behring's Straits, Performed in the Years 1825, 26, 27 and 28*. London: W. Wright, 1836.

ʻĪʻī, John Papa, et al. *Fragments of Hawaiian History*. Honolulu: Bishop Museum Press, 1983.

Macrae, James. *With Lord Byron at the Sandwich Islands in 1825: being extracts from the MS diary of James Macrae, Scottish botanist.* (William Wilson, Ed.) N.p.: Honolulu: 1922

Kaeppler, Adrienne. "'L'Aigle' and HMS 'Blonde': The Use of History in the Study of Ethnography." *Hawaiian Journal of History* 12 (1978): 28–44.

Kamakau, Samuel. *Ruling Chiefs of Hawaii.* Rev. ed, Kamehameha Schools Press: Honolulu, 1992.

Kotzebue, Otto von, and Johann Friedrich Eschscholtz. *A New Voyage Round the World in the Years 1823, 24, 25, and 26.* London: H. Colburn & R. Bentley, 1830.

Shulman, Stanford, Deborah Shulman, and Ronald Sims. "The Tragic 1824 Journey of the Hawaiian King and Queen to London: History of Measles in Hawaii." *The Pediatric Infectious Disease Journal* 28.8 (2009) 728–733. doi:10.1097/INF.0b013e31819c9720

St. John, Harold. "Plants of the Sandwich Islands Collected by James Macrae." *Phytologia* 39 (1978): 307–319.

Stewart, Charles S., and William Ellis. *Journal of a Residence in the Sandwich Islands, during the Years 1823, 1824, and 1825: Including Remarks on the Manners and Customs of the Inhabitants; an Account of Lord Byron's Visit in H.M.S.* Blonde*; and a Description of the Ceremonies Observed at the Interment of the Late King and Queen in the Island of Oahu. By C.S. Stewart, Late American Missionary at the Sandwich Islands: With an Introduction, and Occasional Notes, by William Ellis.* H. Fisher, Son, and Jackson: London, 1828.

Stojanowski, Christopher, and William Duncan (Eds.). *Studies in Forensic Biohistory: Anthropological Perspectives.* Cambridge: Cambridge University Press, 2017. doi:10.1017/9781139683531

"Temperance Complication." *Polynesian*, 9 Nov. 1844: 3.

Thrum, Thomas. *Hawaiian Almanac and Annual for 1923.* N.p.: Honolulu, 1922

Index

The index is based on the terms used in Macrae's original journal. Alternative spellings and related terms, including modern usage and those from the extract and footnotes, are cross-referenced. Persons or places unnamed in the text are indicated, along with the name, when known. Page numbers in **bold** are references to images.

About the Editor

Originally from Canada, Brian Richardson is a graduate of the Political Science and Philosophy Departments at the University of Hawaiʻi at Mānoa. His fields of research include 18th century European exploration in the Pacific, European political philosophy, and information systems in higher education. He has written a book on the voyages of Captain Cook, *Longitude and Empire: How Captain Cook's Voyages Changed the World.*